The Soul *of* DNA

The True Story of a Catholic Sister and Her Role in the
Greatest Scientific Discovery of the Twentieth Century

Jun Tsuji

Llumina Press

The Soul of DNA: The True Story of a Catholic Sister and Her Role in the Greatest Scientific Discovery of the Twentieth Century.

ISBN: 1-59526-206-7
Printed in the United States of America by Llumina Press

In memory of M².

Contents

Preface

*T*he *Soul of DNA* was written for the general reader who possesses some knowledge of high school science. For those who are unfamiliar with basic chemistry or may need a brief refresher, I hope the following explanations will aid you in understanding the chemical structures that appear within these pages.

All forms of matter, such as solids, liquids, and gases, are composed of atoms. There are over one hundred different kinds of atoms, each having a name and symbol; for example, carbon (C), hydrogen (H), oxygen (O), nitrogen (N), and phosphorus (P). Most atoms do not exist in nature as single entities, but instead form chemical bonds with other atoms. Similar to atoms, the different chemical bonds also have names and symbols, such as the single covalent bond (–), double bond (=), and hydrogen bond (...). When two or more different atoms are bonded together, a compound is formed. Compounds are often represented by structural formulas, such as H–H for hydrogen gas or O=O for oxygen gas. For large compounds consisting of many carbon and hydrogen atoms, like thymine shown below, scientists prefer to use a shorthand notation.

Thymine

Instead of using (C) to represent carbon, carbon atoms are implied at the end of lines and at points where lines intersect, such as between the two nitrogen atoms and the lower, right-hand oxygen atom. Also, hydrogen atoms that are bonded to carbons are not shown by this short-hand method.

For all of the chemical structures depicted in this book, it is not important to understand each and every kind of atom, but rather to appreciate the differences in the type and location of the atoms. While it may be difficult to comprehend the atom, because it is too small to be seen, as in religion, you must have faith.

Chapter 1

The DNA Labyrinth

Everything the power of the world does is done in a circle.

– Black Elk

Behind the Motherhouse of the Adrian Dominican Sisters in Michigan there is an eleven-circle labyrinth, a winding path used by worshippers to make a symbolic pilgrimage. The labyrinth, similar to the one on the floor of the nave of Chartres cathedral in France, is a large circle of bricks embedded within crushed black stones, which form distinct lines like those of a maze. In the outer circle, there is an opening where one enters the labyrinth, but unlike a maze, there is only one path to follow. Initially the path leads you straight into the labyrinth, but then directs you along an arc that hairpin turns into the second of a total of thirty-one arcs. As you follow the serpentine path, which winds around the entire surface of the labyrinth, you soon begin to realize how deceptively long and mentally tiresome the journey is. Eventually, the path ends, not at the exit, but in the center of the labyrinth.

Like a labyrinth, the path to discovering the DNA double helix, the three-dimensional structure of DNA, was a long and circuitous journey that ultimately led to DNA's core. DNA, an acronym for deoxyribonucleic acid, is the chemical material that through the processes of evolution, heredity, and development, forms the roots of who we are. During evolution, DNA functions as a record of changes that distinguishes us from other species. DNA also works as a hereditary unit through the replication and transmission of genetic information from one generation to the next. Furthermore, during the course of development, DNA sets the limits of our individual potential, with the environment determining the outcome that is actually realized. Thus, DNA not only determines our history, but also our beginnings as individuals.

For lack of knowledge of DNA, scientists were unable to understand the basis of non-contagious, inborn diseases like cancer – a group of diseases characterized by cells that divide in an uncontrolled fashion. In the past, a myriad of hypotheses have been proposed to explain the cause of cancer. Some of these have included mental misery, the misalignment of divine energy, and the accumulation of natural toxins. Misguided hypotheses, such as these, spurred the development of numerous ineffective, and in some cases, lethal treatments. As a consequence, untold millions have suffered and died of cancer, hopelessly.

The active quest to understand cancer persisted for centuries and eventually led scientists to the DNA labyrinth. The first steps towards revealing DNA were taken by prehistoric humans, who observed that "like begets like" and drew their understanding on the cave walls in southern France. Larger strides were made much later by Gregor Mendel, an Augustinian monk, whose studies of the garden pea provided the basis for our current understanding of heredity through the replication and transmission of genes. After the finding that genes were composed of DNA, scientists were poised to take the final steps in uncovering the secrets within DNA.

The final steps in the discovery of the DNA double helix, however, proved to be the most problematic – scientists were impeded by the shape of the DNA bases. In the early 1950s, the experimental evidence suggested that DNA was a helix composed of two to four chains of sugars and phosphates, to which four kinds of DNA bases were attached. Although the sugars and phosphates formed a regular, repeating unit, the shapes of the DNA bases were irregular. Two of the DNA bases, adenine and guanine, were much larger than the other two, cytosine and thymine, and scientists were unable to conceive of how irregularly-shaped DNA bases could be evenly packed within a helix.

Long after the chemical identification of the DNA bases, scientists debated about the main structural forms of these compounds. At that time, the bases of DNA were thought to co-exist in different configurations, called tautomers – structures that vary in the position of a hydrogen atom and a double bond and spontaneously interconvert.

Thymine, for example, was believed to exist in both the "keto" and "enol" configurations. Likewise, adenine was thought to co-exist in the "amino" and "imino" tautomeric forms. Faced with multiple configurations for each of the four DNA bases, scientists mistakenly proposed structures for DNA that were, in effect, inside-out.

Thymine (keto)   **Thymine (enol)**

Adenine (amino)   **Adenine (imino)**

Then in 1953, James Watson and Francis Crick published their double helix model of DNA and suggested a solution to the problem of the DNA bases. Their DNA model consisted of two helical chains of sugars and phosphates held together in the center by hydrogen bonding between the DNA bases, with the sequence of the DNA bases determining genetic information. In such a manner, the DNA bases formed both the structural and functional "soul" of DNA.

When Watson and Crick proposed the DNA double helix, the predominant tautomeric forms of the DNA bases had not yet been established. Watson proposed the double helix model using the assumption that the most abundant tautomers of the DNA bases were the keto and amino forms. In these configurations, adenine could hydrogen bond with thymine, while guanine could hydrogen bond with cytosine. However, at that time there was little experimental evidence to support that assumption. If the predominant form of the DNA bases was the enol and/or imino tautomer, then hydrogen bonding could not take place between the DNA bases as proposed, which would disprove the double helix model of DNA.

Adenine (amino)   **Thymine (keto)**

Guanine (amino)   **Cytosine (keto)**

Resolving the controversy surrounding the structures of the DNA bases required the application of a chemical identification method known as infrared spectroscopy. Infrared spectroscopy is a technique of characterizing chemical compounds based on their absorption of infrared radiation, wavelengths of light lying just beyond the red limit of the visible spectrum. By this method, chemical compounds are exposed to a source of infrared radiation, and the wavelengths of infrared light that are transmitted and absorbed by the compound are then measured.

Since different combination of atoms, such as those in an enol and keto, absorb different wavelengths of infrared light, this information can then be used as a basis for identification, similar to a fingerprint.

In the early 1950s, it was difficult to prepare biological samples, like DNA, for analysis by infrared spectroscopy. In those days, samples were usually prepared for analysis by mixing them with a liquid to form a solution. However, many biological compounds are not soluble in the liquids commonly used for infrared spectroscopy, and the liquids themselves absorb infrared radiation, which can mask the absorption by the sample compound. These problems were circumvented by the development of a solid preparation technique by Sister Miriam Michael Stimson, OP (Order of Preachers), an Adrian Dominican sister and chemist.

Sister Miriam Michael Stimson, OP, a woman of simple birth and great intelligence, was an anachronism. While following in the medieval traditions of the Adrian Dominicans, Sister Miriam Michael conducted cutting-edge research in her 20th century science laboratory. Guided by her belief in scientific research as an act of worship, Sister devoted her scientific career to the study of the DNA bases.

Isolated from the larger scientific community and parried by male chemists, Sister Miriam Michael, nevertheless, succeeded in developing a solid preparation technique that allowed biological compounds, like the DNA bases, to be analyzed by infrared spectroscopy. The technique she pioneered involved mixing the sample compound with potassium bromide (KBr) powder, which is transparent to infrared light, and then pressing the two in a die to form a translucent disk. The infrared absorption spectrum of the compound could then be obtained from the solid disk. The KBr disk technique that Sister developed was used to provide infrared evidence for the keto and amino tautomers of cytosine and guanine, and the amino form of adenine. These results, along with the data obtained from ultraviolet spectroscopy, x-ray crystallography, and nuclear magnetic resonance studies, confirmed the keto and amino forms as the predominant tautomers of the DNA bases. Hence, Watson's assumption that the DNA bases occurred in the keto, rather than the enol configuration, was affirmed, and in 1962, James Watson, Francis Crick, and Maurice Wilkins were awarded the Nobel Prize.

Knowledge of the structure of DNA then led scientists to a multitude of other discoveries, and subsequently, to a new world of understanding. As a direct result of the DNA double helix, scientists were able to comprehend how genes worked and how mutations could lead to genetic diseases, such as cancer. Knowledge of the genetic basis of cancer has since revolutionized the way cancers are diagnosed and treated.

All these steps, and more, paved within the circles of life. To walk the DNA labyrinth – to retrace the path that led to the discovery of the double helix – is to see anew, the soul of DNA.

Chapter 2

A Disease of Our Genes

Hopeless in the eyes of the patient;
hopeless in the eyes of the surgeon;
afraid even to use the word.

– Charles Chide, 1906

*I*n the summer of 1885, President Ulysses S. Grant, one of America's most revered war heroes, lay dead following an exacerbating illness caused by, according to one doctor, an "inflammation of the epithelial membrane of the mouth."[1] While Grant's physicians were well aware that the malignancy that had engulfed his mouth and throat was cancer, many were reluctant to speak the word – for many doctors were humbled by their own lack of understanding of cancer and their inability to control this devastating disease. Recognizing this and the advanced state of Grant's cancer, his physicians conceded. There was nothing that could be done, except to reduce his suffering with painkillers.

The widely publicized death of President Grant is representative of the state of affairs, prior to the twentieth century, concerning the cause and treatment of cancer. Accounts of such deaths, although not often discussed in public, were not new. Hundreds of thousands of people have lost loved ones to cancer. This disease leaves an indelible impression in the memory of those who have witnessed the emaciated, cadaverous face of an advanced cancer patient and the emotional trauma it inflicts on family members and friends. Even the word "cancer" conjures perdition-like images of a slow, painful, fatalistic death devoid of any hope – making "cancer" one of the most feared words. The countless deaths and copious tears are directly attributable to the absence of effective methods of treatment, which stem from a lack of understanding of cancer.

Prior to the middle of the twentieth century, the fight against cancer was in a state of disarray. Epidemic numbers of people were dying of cancer, yet no one really understood its roots. As a consequence, traditional medical care only involved treating the symptoms, often by surgery, rather than correcting the underlying problem. Surgery had never been a welcome option to patients, in part, because of its macabre reputation. For instance, of 170 mastectomies performed to remove breast tumors between 1867 and 1876, only 4.7 percent of the women were still alive three years after the surgery.[2] The ineffectiveness of surgery was realized even by the ancient Greek physician Hippocrates, who aptly advised, "It is better not to apply any treatment in the case of occult [hidden] cancer; for if treated, the patients die quickly; but if not treated, they hold out for a long time."[3]

Faced with such prospects, many cancer patients turned to home remedies or "quacks" for help. Many of these "alternative" methods involved setting one malignant force against another. To some, the remedy involved placing crow's feet, brimstone, arsenic, or even a dead person's hand on the site of the tumor. One "cure" even advised drinking a concoction of boiled crayfish in asses' milk. Another "cure-all" entailed walking backwards with one's clothes inside-out, while simultaneously throwing things over the shoulder and cutting one's fingernails on a Friday during a waning moon.[4] While such anecdotes invite laughter today, they exemplify the erroneous perceptions surrounding the origins of cancer and its treatment.

Faced with a paucity of accurate information, even the learned community put forth countless cancer "theories" and "cures." The origins of many of these ideas can be traced to ancient Egyptian, Greek, and Roman healing practices, many of which were based on principles of holistic medicine. Mind, body, and spirit were viewed as interconnected, and an imbalance in one was believed to cause problems in another. The search for the causes of these imbalances gradually led researchers to the environment. Drinking water, pollutants, and pathogenic organisms were each imputed as the roots of cancer. Eventually, attention would return to the body, as not only the site of the affliction, but also the source of the symptoms.

One of the earliest known "theories" concerning the origins of cancer involved an imaginary body fluid called "black bile." Black bile was one of four types of liquids that the ancient Greeks and Romans believed were contained in harmony within the body. They believed that an imbalance of one could result in diseases, like cancer. When cancerous bodies were dissected and examined, the ancient Greek physician Hippocrates, and later Cladius Galen, observed bloody swellings of tissue, which they attributed to an excess of black bile. They believed that the flow of excessive black bile explained how cancer could spread within the body. Hippocrates and Galen also coined the terms "carcinoma" and "cancer," which mean "crab" in Greek and Latin, respectively, because of the likeness between the extended veins radiating from a tumor and the legs of a crab. But what did they believe caused too much black bile? The answer lies in the word itself, for "black bile" is the English translation of the Greek word for "melancholy."[5]

The ancient Greeks and Romans were among the first to suggest that problems of the mind can cause problems in the body. In keeping with such views, they believed that mental misery was the underlying cause of cancer. Hippocrates and Galen, for example, both viewed breast cancer as the result of depression, a view that would continue even into the nineteenth century. Proponents of this idea believed that emotional stress, for instance, caused by the loss of a close relative or by financial difficulties, induced a bodily imbalance giving rise to cancer. Such views were also invoked to explain the proclivity of women, who were seen as emotionally more unstable than men, to develop breast cancer. Currently, there is no credible evidence to support such assertions, but in the past, many also believed that the problems of mental depression were related to the psyche, the unity of mind and soul, and to one's connection with a spiritual being.

The ancient Greeks also believed that problems of the spirit can affect the body. Since the ancient Greeks, many have believed in the existence of an energy force that connects humans with a supreme being. Diseases, like cancer, were thought to form when a person's energy field became misaligned with that of a divine-being. Ancient cultures also believed that crystals possessed supernatural powers that harbored spirits as well as emitted light. They believed that certain crystals, like quartz, malachite, and other gemstones, possessed special energy that could heal underlying problems. Treatments often involved locating the site of the energy "blockage" and then placing particular crystals on specific parts of the body. Different crystals were thought to contain different healing powers – red crystals were used to treat blood problems, blue crystals were used for the brain,

and so on. Despite any scientific evidence of any special healing powers, the use of crystals, and the holistic view of the body-spirit
connection remain popular.

While the ancient Greek and Roman physicians distrusted the mind
and spirit, the ancient Egyptians suspected the body as the source of the
cancer problem. Similar to the holistic views of health held by the
Greeks and Romans, the Egyptians believed that problems of one body
part could cause problems in others. Since the ancient Egyptians, many
have believed that diseases, including cancer, occurred due to the accumulation of toxic waste products in the body. Proponents claimed
that stagnation causes natural toxins to form, which poison the body
and cause disease. Various cleansing and detoxification "cures" have
been proposed including fasting, bloodletting, enemas, and even colon
purges. The latter method was very popular in the United States in the
1920s and 1930s, when flushing the colon was thought to prevent the
accumulation of harmful waste. Fasting was also believed to cleanse
the body by eliminating the intake of foods – the source of the toxins.
However, there is no scientific data that any of these are effective cancer treatments, and contrary to claims, these methods can be dangerous
and can possibly lead to death. These methods do, however, highlight
the problems associated with developing effective cancer treatments
when the cause of cancer is uncertain.

By the mid 1800s, with the advent of the perfected microscope, attention returned to the tumor itself. The early microscopists discovered
that tumors were composed, not of black bile, but of cells, like those
found in other parts of the body. Normally, the human body makes new
cells by a process called cell division, where cells split to form more
cells. This process is essential for normal growth and development. For
instance, more cells are necessary for a single fertilized egg cell to
grow and develop into a multi-cellular adult. Also, cells do not live in-
definitely, and new cells are required to replace old or damaged ones,
such as the cells of the intestinal lining, which multiply daily. When
you accidentally cut or scrape your skin, new cells from your inner skin
layer are made to heal the wound. Still others, like the white blood cells
of your immune system, reproduce in response to foreign bacteria or
viruses. Usually, cells grow and divide in an orderly manner and only
when and where they are needed. A different situation occurs with cancer cells.

With the aid of the microscope, scientists began to understand cancer as the result of uncontrolled cell division. When viewed under the
magnifying power of lenses, the cancerous cells were found to "have

lost their characteristic structure, polarity, and arrangement. They pay
no heed to limiting supporting membranes but invade them to continue
multiplying wherever they obtain sufficient nutrition for so doing."[6]
Normally, cells stop growing when they come in contact with one another and form highly ordered arrangements. Cancer cells, on the other
hand, were found not to be inhibited by such contact and formed piles
leading to the disordered array of cells seen in tumors. Early microscopists also noticed that most cancer cells possessed abnormal
chromosomes, DNA- and protein-containing intracellular structures.
The microscope enabled scientists to understand cancer in an entirely
new perspective, which firmly established cancer as a cellular disease.

For many years, cancer researchers debated whether cancer was a
single disease or a set of diseases. This controversy was partly settled
through the results obtained from cancer transplantation experiments.
In the late 1800s, scientists observed that when cancerous rat cells were
transplanted to healthy rats, tumors would develop on the normally
healthy rats. These experimental findings revealed that cancer cells are
autonomous – they grow by their own division, not by converting
neighboring cells into cancer cells, and they are independent of the
rules that govern the growth of neighboring tissues. Cancer researchers
also transplanted cancerous cells to entirely different tissues. For instance, when cancerous liver cells were transplanted to the brain, the
cancerous cells continued to grow and function as liver cells, not brain
cells. The results of these experiments demonstrated that each type of
cancer is determined by its tissue of origin and can behave differently
from cancers established in other locations of the body. Thus, cancer
became viewed as a collection of diseases, whose symptoms and treatments would vary depending upon the type of cancer.

The American Cancer Society collects annual statistics on nearly 50
different types of cancers.[7] Each type of cancer is classified by its tissue
of origin; for example, cancers that begin on the skin are referred to as
skin cancers. The cancer statistics are similar between men and
women; however, there are some important differences. Among men,
the greatest number of new cancer cases are of the genital system (pros-
tate), followed next by the digestive system (colon) and then by the
respiratory system (lung). In contrast, women will be diagnosed with
breast cancer more than with any other type, followed next by cancers
of the digestive system (colon) and then of the respiratory system
(lung). Among both sexes, lung cancer claims more lives than any other
type of cancer.

	Estimated Number of New Cancer Cases in 2002	
Type of Cancer	Males	Females
Genital (men) or Breast (women)	197,700	203,500
Digestive System	130,300	120,300
Respiratory System	100,700	82,500

	Estimated Number of Cancer Deaths in 2002	
Type of Cancer	Males	Females
Respiratory System (lung)	94,100	67,300
Digestive System (colon)	70,800	61,500
Genital (men) or Breast (women)	30,900	39,600

By the beginning of the twentieth century, the more prominent cancer "theories" focused on the environment, rather than the mind, body, or spirit. One of the more bizarre of these hypotheses was the notion that drinking trout-inhabited waters caused cancer. In the early 1900s, H. R. Gaylord, director of the New York State Institute for the Study of Malignant Diseases, noticed an "astonishing coincidence" between the occurrence of cancer and the distribution of trout within the state of New York. He wrote, "A map of one might well be taken as a map of the other." He later claimed that 100 million of the state's trout had cancer and that people who drank from trout-filled waters ran the risk of developing the disease. Following Gaylord's advice, President William Taft recommended that Congress appropriate $50,000 to study this issue. The President's proposal was well received by the media, and it was described by *The New York Times* as "the greatest stroke so far toward the conquest of the dread disease." Although the idea appealed to the public's imagination, other scientists did not share in Gaylord's fear of trout, and the proposal was dismissed by the Committee on Fisheries.[8]

Despite the trout debacle, many scientists continued to believe that environmental factors were at the roots of cancer. One of the more popular of these ideas was the "chronic irritation" hypothesis. According to this idea, long-term exposure to an irritating substance, such as an environmental pollutant, was thought to aggravate cells and make them grow out of control. Evidence for this hypothesis originally came from numerous observations of occupation-related cancers, like cancer of the scrotum among chimney sweeps, lung cancer in miners, and skin

cancer among those who worked with radium and x-rays. Such observations led many scientists to search for specific irritants, and in 1915,
two Japanese scientists, Yamagiwa and Ichikawa, discovered that skin
cancers could be induced on rabbits by the long-term application of
coal tar on their ears. The discovery of additional irritants soon followed; however, no substance has received more attention than
tobacco.

The widely publicized death of President Ulysses S. Grant, a heavy
cigar smoker, of mouth cancer in 1885, did not deter future generations
from smoking tobacco. In America, the rise in the popularity of smoking began shortly after World War I, upon the return of the doughboys
with their nicotine-dependency. By 1925, half of all adult men, and a
growing number of women, were smoking.[9] While the death rate due to
lung cancer was nominal at that time, by the 1950s, men would die of
lung cancer more than from any other form of this disease. Likewise,
by the 1990s, lung cancer would claim the lives of more women than
breast cancer.[10] By the 1950s, epidemiologists had firmly established a
correlation between cigarette smoking and lung cancer. However, not
everyone who smoked developed cancer, and women were more prone
to lung cancer even though fewer women smoked than men. Detractors
were quick to point out such observations, which cast a cloud of suspicion on the role of irritants and cancer.

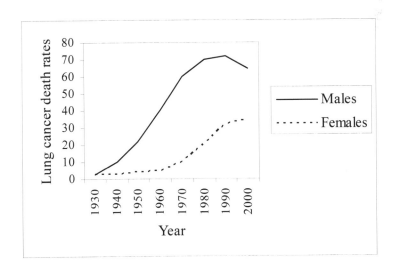

Although the chronic irritation hypothesis has been supplanted by
more contemporary hypotheses, this body of work helped to demonstrate the importance of environmental risk factors. A risk factor is

anything that increases a person's chance of developing cancer. A recent study of identical twins demonstrated that environmental factors account for over half of all cancers.[11] The most significant of these factors include cigarette smoking, an unhealthy diet and lack of exercise, and exposure to excessive sunlight, which are associated with lung cancer, cancer of the colon and pancreas, and skin cancer, respectively.

Another environment-related cancer hypothesis that was widely believed in the past was the "infectious agent" hypothesis. This idea grew out of the nineteenth century, when the causal agents of many infectious diseases were discovered, and many believed cancer would yield similar results. The infectious agent hypothesis promulgated the idea that an infectious entity from the environment, like a pathogenic bacterium, could invade cells and induce them to grow in an uncontrolled manner. The notion that cancer was contagious was widespread among the public. Cancer hospitals and wards were created so cancer patients could be quarantined from the public to prevent the spread of disease. The clothes and belongings of cancer victims were also commonly burned for fear that they may be infectious. The idea that cancers arose from bacterial or viral infections seemed promising when biologist Peyton Rous discovered in 1910 that a cell-free filtrate of macerated chicken tumors could induce cancers when injected into healthy chickens. This result suggested that some kind of filterable agent, like a virus, was the cause of cancer. However, Rous and others were unable to find viruses associated with human cancers. Furthermore, there was little scientific evidence that cancers were contagious, since spouses, doctors, and nurses of cancer victims did not have higher incidences of cancer. For these reasons, the infectious agent hypothesis lost favor among scientists.

Cancer was also imputed to be entirely genetic, a notion that promoted fatalism. Cancer was said to be in your blood and there was nothing you could do about it. Because of the higher frequency of a few types of cancers, like breast cancer, among certain families, cancer was also seen by many as a familial disease. Cancer, like criminality, was considered a family stigma and was often cloaked in secrecy. During the eugenics movement of the early 1900s, many even avoided marrying men and women from such families for the fear of having children with cancer.

Some of the earliest experiments on the genetics of cancer were performed with mice by the American biologist Maud Slye. In the early 1900s, she bred two types of mice – a cancer strain that would develop breast cancer, and a non-cancer strain that was resistant to breast cancer. By mating these two mice strains and observing the offspring, she discovered that susceptibility to breast cancer was inherited. She also studied other cancers, mating over a 100,000 mice, and found that all types of mouse cancers have some genetic basis. Her pioneering work established the idea of a hereditary predisposition for developing cancer. In other words, some people are genetically more prone than others to certain forms of cancers.

The view that genes were the basis for cancer proved to be a polemical idea in the first half of the twentieth century. There were many reasons for the slow acceptance of the hereditary basis of cancer. First, the genetic evidence was not unequivocal. For example, many descendants of "cancer families" did not develop cancer. Likewise, cancer occurred in many individuals who were born from "cancer-free families." Secondly, most scientists favored either the chronic irritation or the infectious agent hypothesis, and they did not see the relationship between irritants, viruses, and genes. Lastly, many viewed genes as metaphysical entities that were convenient explanatory symbols, but did not actually exist. Until scientists could understand what a gene was, the genetic theory of cancer would remain controversial.

Chapter 3

The Roots of Humanity

Being is eternal; for laws there are to conserve the treasures of life on which the Universe draws for beauty.

– Goethe

For centuries, humans have been cognizant of the inheritance patterns exhibited by wild animals and plants. One of the earliest examples of this understanding is demonstrated in the art left by prehistoric humans in the Grotte de Niaux in southern France approximately 10,750 B.C. The walls of the cave of Niaux are decorated with drawings of adult and young bison, which demonstrate that the early humans not only realized that bison give birth to bison, but also that the offspring share the same characteristics, such as rearward-curved horns, as the adults. By 6000 B.C., the ancient humans also understood that "like begets like" applies to plants, as evinced by the cultivation of hybrid wheat at Jericho. In this ancient Middle-Eastern city, archeologists have unearthed primitive sickles and planting devices, demonstrating that the early humans understood that sowing a wheat seed will produce a plant with the same valuable attributes, such as the potential to make bread flour, as the plant from which the seed was harvested. These realizations would eventually lead to the widespread cultivation of plants and the domestication of animals.

With the advent of agriculture, the early farmers further applied their understanding of heredity by artificially selecting and selectively breeding animals and plants. The early agriculturalists recognized that variation existed within their herds – some animals grew larger than others and were sources for more meat, hide, milk, or wool. Likewise, the early farmers observed differences in the size, shape, and taste of their crops. The early humans then began to artificially select the most desirable animals and plants. For example, from a litter of semi-tamed wolves, the best companions and most alert sentries were retained, which would eventually give rise to different breeds of dogs, while the more aggressive ones were consumed for dinner. Similarly, plants were artificially selected by storing seeds of the most desirable plants for sowing in the next growing season, while eating the others. The early farmers also selectively bred their animals. By choosing the most desirable animals and separating them from the others during the mating season, the early humans were able to control the types of attributes that would be passed on to the next generation. Similarly, the early farmers also practiced selective plant breeding. Through the recognition of the male and female reproductive organs of plants, the early farmers were able to control inheritance by artificially pollinating the most desirable plants. Through artificial selection and selective breeding practices, new animal breeds and plant varieties were created, which in turn provided more food, fibers, and hides.

The application of their rudimentary understanding of heredity had many important consequences for the ancient humans. Prior to agriculture, humans struggled for survival as nomads. This harsh existence consisted of following the seasonal migration of wild herds, their primary food source, and scavenging for wild fruits and seeds. Hunting, gathering, and preparing food consumed all of their time and energy. Because of their peregrinations, there was little if any time for innovations, such as written language, art, or even new thoughts. The use of genetics, in the form of agriculture, changed all that. According to science historian Jacob Bronowski, agriculture was "the largest single step in the ascent of man"[1] and led to the birth of civilization. Through the cultivation of plants and the domestication of animals, farmers were able to create a more successful and reliable method of obtaining food. Without the need to expend great lengths of time and energy scavenging for food and securing shelter, the early farmers could now erect permanent structures, develop written language, and invent new devices. They also had more time to ponder their world and to contemplate their own origins: Why do dogs always give birth to dogs, goats to goats, and wheat to wheat? Why do I look like my parents?

While many great thinkers have attempted to understand the nature of heredity, Aristotle was among the first to answer these questions with a modicum of accuracy. In the third century B.C., he propounded the idea that each part of every new organism is found in the menstrual blood of its mother and is activated by the semen of its father. This idea would later develop into the concept of preformation, the belief that each offspring inherits all of its features from only one parent. Most preformationists believed that a fully formed, miniature human, called a homunculus, resided within each sperm cell. Presumably, each male homunculus also had sperm with their own preformed, miniature humans inside, which in turn had their own homunculi. They viewed each person as like a series of nested eggs consisting of thousands of progressively smaller miniaturized humans. For these preformationists, the womb was simply a fertile ground within which the embryo would grow from the seed-like sperm. Consequently, they believed that the father was the sole contributor to a child's inherited characteristics. On the other hand, a few preformationists believed that the homunculus resided in the egg and that the sperm merely activated the growth of the preformed, miniature human. In this case, the child was thought to inherit all features from only the mother. So pervasive was the belief in preformation that an early microscopist even imagined that he could see a homunculus! Today, scientists describe the homunculus as a myth for there is no evidence that tiny, coiled-up humans reside in sperm or eggs. Preformation has also been discredited by the scientific finding that both the mother and father contribute towards the attributes of their children.

When looking at a child, one can often see a combination of the parent's features, an observation which may have contributed to the concept of blending inheritance. Blending inheritance is the belief that parental traits mix and cause offspring to take on an intermediate form different from that which is present in either parent, similar to the way that blue and yellow paints merge to form green. For example, when a red-flowered four-o'clock plant is crossed with a white-flowered variety, a hybrid is created that makes pink flowers – an observation that superficially supports the concept of blending inheritance. However, when the pink-flowered hybrids are crossed with one another, the resulting offspring do not make pink flowers as predicted by the blending concept. Instead, some of the offspring produce red flowers, while others make white flowers, indicating the mendacity of blending inheritance. Why some traits, like red flower pigment, sometime disappear and then mysteriously reappear in subsequent generations puzzled breeders for centuries. Not until the middle of the nineteenth century would the experiments of a Catholic monk, named Gregor Mendel, shed light on the basic laws governing inheritance.

Gregor Mendel was an Augustinian monk of the Order of St. Thomas in Brno who is credited with the discoveries that laid the foundations for our current understanding of heredity. Having failed his university exams and unable to qualify as a teacher, Mendel decided to devote himself to the study of plant inheritance in the gardens of the monastery. Using pure-bred varieties of the garden pea, Mendel performed a series of monohybrid crosses – matings between plants that differ in only one trait. For instance, Mendel transferred pollen from a yellow-seeded variety to the flower of a green-seeded plant and then observed the color of the seeds produced by the resulting F1 offspring. Mendel observed, in contrast to the intermediate appearance predicted by the blending hypothesis, that the resulting hybrids produced only yellow seeds. Mendel then allowed the hybrids to self-fertilize and repeated his analysis with the next F2 generation. He observed that approximately three-fourths of the F2 plants made yellow seeds, while the remainder produced green seeds. Based on the results of his monohybrid crosses, Mendel correctly surmised that each trait is governed by discrete units, rather than amorphous liquids that could be blended together, which he called "elemente." Mendel also suggested that the hereditary units occurred in different forms, which are known today as alleles. Mendel envisioned that every pea plant carried two alleles for each hereditary unit, having received one from its maternal parent and the other from its paternal parent. The concept that offspring inherit distinct hereditary units from each parent, without being blended, would later be known as the particulate theory of inheritance.

Two hereditary "laws" arose as a result of Mendel's work. The first, known as the principle of segregation, explains how alleles are transmitted from parent to offspring. Mendel invented a system of symbols that helped him to reach this understanding. He represented each trait with a letter, using an uppercase letter (A) to symbolize the dominant allele, the form expressed in the F1 hybrid, and the lower- case letter (a) for the recessive allele, the form masked by the dominant allele in the F1 hybrid. Mendel correctly deduced that the two alleles for each trait separate during the formation of gametes (egg and sperm). For example, Mendel's pure-bred, yellow-seeded variety (AA) would produce sperm-bearing pollen grains that contain only the (A) allele for pea color, while the pure-bred, green-seeded variety (aa) would make egg-bearing flowers with only the (a) allele. The fusion of these gametes would then result in an F1 hybrid containing two alleles (Aa). Upon self-fertilization, the F1 hybrids would make two types of gametes – half carrying (A) and the other half possessing (a). Mendel correctly surmised that these gametes would unite randomly at fertilization to produce all possible combinations. The resulting F2 generation would then consist of one-fourth of each of the following combinations: (AA), (Aa), (aA), and (aa). The concept that, for each trait, the two alleles segregate during the creation of egg and sperm and unite randomly at fertilization would later be known as Mendel's first law of heredity.

A diagrammatical representation of one of Mendel's monohybrid crosses.

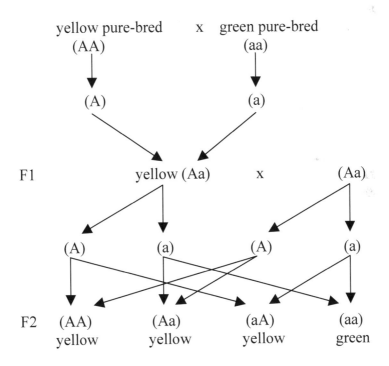

Mendel's second "law" of heredity arose as a result of his dihybrid crosses – matings between plants that differ in two unrelated traits. In one such example, Mendel examined the inheritance of pea color (yellow or green) and seed texture (round or wrinkled) within the same plants. For instance, Mendel crossed a pure-bred, yellow- and round-seeded variety (AABB) with a pure-bred, green- and wrinkled-seeded plant (aabb). From this cross, he obtained F1 hybrids (AaBb) that produced yellow, round seeds. Regarding this hybrid, Mendel was interested in determining how the alleles of the two traits would segregate during the formation of its gametes. One possibility is that the dominant alleles (AB) could segregate together, and the two recessive alleles (ab) could do the same. Alternatively, the alleles of the two traits could assort independently of one another and produce all combinations of dominant and recessive alleles (AB, Ab, aB, and ab). To assess these two possibilities, Mendel crossed the F1 hybrid (AaBb) back to its green- and wrinkled-seeded parent (aabb). He observed that the resulting offspring produced roughly equal numbers of seeds that were yellow and round (AABB), yellow and wrinkled (Aabb), green and round (aaBb), and green and wrinkled (aabb) – results that supported the second of the two possibilities. These results, among others, confirmed that the alleles of the two traits were assorting independently of one another, which became the basis of Mendel's second law of heredity. In 1865, Mendel published the results of his inheritance study of the garden pea, which went largely ignored until after his death.

A representation of one of Mendel's dihybrid crosses.

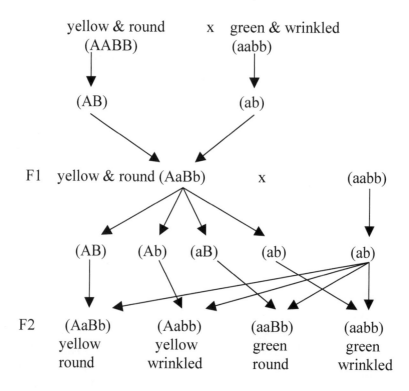

In the years following the rediscovery and independent confirmation of Mendel's work, two terms were introduced that have become part of our vernacular language. "Genetics" was coined in 1906 by English zoologist William Bateson, who used the term to describe the new understanding of inheritance that had replaced the blending concept and preformation. Three years later, Danish botanist Wilhelm Johannsen, perhaps unaware of Mendel's term "elemente," saw the need for a new term to replace those associated with the blending concept, and introduced the word "gene." Johannsen wrote, "it expresses only the evident fact that, in any case, many characteristics of the organism are specified in the gametes (eggs and sperm) by means of special conditions, foundations, and determiners which are present in unique, separate, and thereby independent ways – in short, precisely what we wish to call genes."[2]

In the following decades, the gene was viewed by many as a metaphysical concept. Many geneticists thought of genes as a practical way of explaining heredity without the need for genes to actually exist. Nobel laureate Barbara McClintock, for example, referred to the gene as "merely a symbol."[3] Others were highly suspicious of what they saw as the overuse of genes as explanatory devices. In the absence of any other explanation, genes were all too often invoked as a convenient explanation for the unknown. However, this view of the gene would gradually change beginning in 1927, when Hermann Muller demonstrated that x-rays were mutagenic, presumably by altering the chemical make up of genes. Muller's experimental findings could be explained only if the gene was a physical entity constructed of atoms, like all other forms of matter.

Meanwhile, with the advent of the perfected microscope, scientists had identified stainable bodies, called chromosomes, inside of cells. The chromosomes were only visible at certain times, such as during the creation of egg and sperm. These cells were known to play important roles in reproduction and were inferred to be the carriers of genetic information between generations. The microscopic observation that the number of chromosomes is halved during the creation of egg and sperm and fully reconstituted in the fertilized egg prompted Walter Sutton in 1903 to propose that the chromosomes "constitute the physical basis of the Mendelian law of heredity."[4] Later in 1910, Thomas Hunt Morgan's genetic studies of the fruit fly revealed that eye color and gender were linked to the presence of the X chromosome. Female flies, for example, were found to possess two X chromosomes, while males only had one, suggesting that the X chromosome contained genes for femaleness. Such work not only provided evidence for the existence of genes on chromosomes, but also further established the concept of genes as physical entities. This body of work also spurred other scientists to investigate the chemical nature of the chromosome.

While scientists could now see chromosomes, several years lapsed before they could determine what the chromosomes were actually made of. In 1923, using a procedure developed by Robert Feulgen, the chromosomes were found to contain DNA. The procedure relies on a chemical called the Schiff reagent, which stains DNA red. In a preparation of cells treated with Feulgen's stain, the chromosomes reddened, while other areas of the cell were relatively colorless. Although a small amount of DNA is present in small energy-producing structures called mitochondria and in photosynthetic chloroplasts, this colorimetric reaction showed that DNA was localized primarily to the chromosomes.

DNA, an acronym for deoxyribonucleic acid, was discovered in 1869 by the Swiss biochemist Friedrich Miescher. Miescher worked with cells obtained from human pus, a fluid often excreted from surgical wounds prior to the widespread use of antiseptics. Miescher extracted a weakly acidic, phosphorous-rich substance from large membrane-bound compartments of these cells, called nuclei, and named the substance "nuclein," which was later renamed "nucleic acid." At that time, living organisms were thought to be constructed of only three large groups of compounds: proteins, carbohydrates, and fats. The substance Miescher had isolated was unlike any of these compounds. While Miescher erroneously believed DNA was a phosphate storage compound, later scientists would discover DNA's true structure and function.

Feulgen's discovery of DNA in the chromosomes did not prove that DNA had a role in heredity. Chromosomes also contain a relatively large amount of protein, assemblages of small building blocks called amino acids. At that time, most scientists viewed DNA as a simple, repetitive molecule that functioned as a scaffold for the proteins. In addition, DNA is made up of a combination of only four types of nucleotides, and these four were thought to be too few to account for all the genetic diversity of life. Proteins, on the other hand, built from different combinations of twenty amino acids, were believed to be more reflective of the kind of variation observed among living organisms. For these reasons, proteins were believed to be the chemical material of genes.

In 1928, English bacteriologist Frederick Griffith's transformation experiments provided the beginnings for our understanding of genes as segments of DNA. Griffith worked with two strains of the pneumonia-causing bacterium *Streptococcus pneumoniae* – a pathogenic S strain, which produced smooth-looking colonies, and a nonpathogenic R strain, colonies of which had a rough appearance. Griffith discovered that mixing the heat-killed S strain with live R strain transformed the nonpathogenic R strain into a pathogenic S strain. Later, it was sur-mised that the transforming substance from the heat-killed S strain was composed of genetic material. In 1943, microbiologists Oswald Avery, Colin MacLeod, and Maclyn McCarty purified the transforming sub-stance from the heat-killed S strain and discovered that it retained its activity even after it had been treated to destroy proteins, carbohy-drates, and fats. Alternatively, the transforming substance lost its activity when treated with an enzyme that destroys DNA. Taken to-gether, these experiments demonstrated that the hereditary material from the S strain was DNA. However, skeptics claimed that the trans-forming substance was contaminated with protein, and it was by no means generally accepted that DNA was the genetic material. Many scientists continued to believe that genes were made of protein.

Confirmation of the idea that genes were made of DNA came in 1952, when Alfred Hershey and Martha Chase published the results of their bacteriophage experiments. Bacteriophages are viruses that are composed of only protein and DNA and infect certain types of bacteria. During the infection process, the virus injects its genetic material inside of the bacterium, which then directs the synthesis of new viruses. To determine whether the viral genetic substance was made of protein or DNA, Hershey and Chase tagged the viral proteins with radioactive sul-fur and the viral DNA with radioactive phosphorus. Following infection, they used a Waring blender to dislodge the viruses from the bacteria, which they then separated. Afterwards, they discovered only radioactive phosphorus inside of the bacteria – evidence that the viral genetic material was made of DNA, not protein.

By the early 1950s, the scientific community finally accepted the notion that genes were made of DNA. Since genes function as the he-redity units responsible for the continuity of life, scientists wanted to know how genes could be replicated and be stably transmitted across generations. Since the function of a molecule is directly related to its structure, determining the structure of DNA became paramount for un-derstanding how genes worked.

Chapter 4

Fibers of Life

*There is nothing over which a free man ponders less than death; his
wisdom is, to mediate not on death but on life.*
— *Baruch Spinoza*

*I*n 1944, Nobel laureate Erwin Schrödinger published a book in
which he challenged the reader to think of biological organisms in
terms of physics and chemistry. In his book entitled *What is Life?*,
Schrödinger used physics to frame questions of biological importance.
How can genes account for both the permanence and the vicissitude of
traits? According to Schrödinger, the answer could be found in the
structure of genes and chromosomes. In contrast to the simple, repeating patterns exhibited by "periodic crystals," to Schrödinger, genes and
chromosomes were "aperiodic crystals" that exhibited "no dull repetition," but were "elaborate, coherent, and meaningful."[1] The "aperiodic"
structure of a gene, Schrödinger argued, could possess more genetic
information than the confined structure of a simple, repeating unit.
Such information was encoded within what Schrödinger called the "hereditary code-script," or what we call the "genetic code" today. The
point is then made that the number of parts of the code can be small,
and yet generate sufficient information to account for the diversity of
life. A genetic code consisting of only five different units arranged in
groups of up to twenty-five, for example, would result in over 370
quadrillion possible combinations![2] Astonished by how a "miniature
code" consisting of a small, but highly organized group of atoms is able
to carry the complete potential to make a whole organism, Schrödinger
concluded his book by poetically referring to the code-script as "the
finest masterpiece ever achieved."[3] *What is Life?* influenced and in-

spired many scientists, especially physicists, to study DNA. Schrödinger's book helped many to see the relationship between the structure of DNA and its function as a genetic code controlling ontogeny. Schrödinger's underlying message was clear – discover the structure of DNA and the inner workings of life will be unveiled.

Before the structure of DNA could be determined, scientists first had to sort out DNA from the mixture of other nucleic acids that were thought to exist at that time. The late 1800s and early 1900s were periods of uncertainty regarding the chemical nature of the nucleic acids. The literature of that time was filled with the names of various types of nucleic acids such as "true", "para-," "pseudo-", "thymus", and "yeast" nucleic acids. Later with the development of improved isolation techniques, "true" nucleic acid was found to consist only of DNA, while para- or pseudo-nucleic acid contained DNA and its associated proteins. Meanwhile, it was believed that a sharp distinction existed between the nucleic acids of animals and plants. The nucleic acids isolated from thymus glands of animals contained the DNA base thymine, while those obtained from yeast and wheat possessed the base uracil. Later, once all the nucleic acid components had been discovered, it was realized that thymus nucleic acid consisted of DNA, which also occurs in the nuclei of plant and yeast cells, while the yeast nucleic acid was composed of RNA, which also exists in the cytoplasm of animal cells. By the end of the 1930s, there was general agreement in the existence of only two types of nucleic acids, DNA and RNA, which are found in both animal and plant cells.

	DNA	RNA
Purines	Adenine and Guanine	Adenine and Guanine
Pyrimidines	Thymine and Cytosine	Uracil and Cytosine
Sugar	Deoxyribose	Ribose
Phosphate	One per sugar	One per sugar
Cellular location	Nucleus	Cytoplasm
Former name	Thymus nucleic acid	Yeast nucleic acid

Following the discovery of the nucleic acid components, much research focused on how the parts were connected. At that time, DNA was known to be composed of phosphates, sugars, and four kinds of nitrogen-containing bases – 2 purines (adenine and guanine) and 2 pyrimidines (thymine and cytosine). However, it was unclear how these pieces were arranged to form DNA. Much of the early proposed structures of the nucleic acids arose from data obtained from the partial or complete hydrolysis of DNA and RNA. The partial breakdown of the nucleic acids was used to determine what parts were connected to others. For instance, it is easier to find the solution to a jigsaw puzzle if many of the pieces are already joined. Likewise, by breaking some, but not all of the bonds that link the nucleic acid components, one could ascertain what pieces remain connected and use that information to infer the complete structure. Alternatively, by completely hydrolyzing DNA, thoroughly breaking it up into its component pieces, scientists could determine the proportions of each part. Such experiments, for example, demonstrated that nucleic acids consisted of equal amounts of sugars and phosphates. In theory, this approach could also be used to determine the ratios of the four types of nitrogen-containing bases. However, in practice, the complete hydrolysis of DNA often chemically modified or destroyed many of the DNA bases, giving inconsistent results that were not reflective of the true ratios of the bases. The hydrolysis of DNA in 1905, for instance, yielded more thymine than adenine and more cytosine than guanine, while the opposite results were reported in the following year. Although faced with conflicting findings, the interpretations of these data by Phoebus Aaron Levene, an English biochemist, would alter the course of nucleic acid research for the next thirty years.

Phoebus Aaron Levene promulgated the idea that nucleic acids were composed of repeating blocks of the same four nucleotides. DNA, for example, was viewed as a simple repetition of thymine, adenine, cytosine, and guanine (i.e. TACG TACG TACG...). This belief would later become known as the tetranucleotide hypothesis. Proponents of the tetranucleotide hypothesis viewed nucleic acids as unsophisticated molecules incapable of significant biological roles. As a consequence, the tetranucleotide hypothesis perpetuated the false belief that the complicated protein, rather than the simple DNA, was the complex genetic substance. Unfortunately, the tetranucleotide hypothesis became dogma and affected the way scientists viewed the structures of DNA until the 1940s.

One of the earliest proposed structures for a nucleic acid was introduced in 1902 by Thomas Osborne and Isaac Harris. From wheat embryos, they isolated RNA, which they called "triticonucleic acid." Upon hydrolysis of this nucleic acid, they observed roughly equal amounts of adenine and guanine, slightly more uracil, and traces of an unidentified base, which may have been cytosine. Based on these results, they proposed a structure for triticonucleic acid, which consisted of sugars and nitrogen-containing bases attached to a phosphate "backbone," where "X" represents the unidentified base. Their irregularly-shaped model, though incorrect, serves to highlight just how far scientists had to progress to arrive at the correct structure.

Then in 1912, geneticist F. P. H. Steudel proposed one of the first models for the structure of DNA. Although unsupported by the results of his DNA hydrolysis experiment of 1906, Steudel favored Levene's tetranucleotide hypothesis and envisioned a structure with equal proportions of adenine, guanine, thymine, and cytosine. Steudel's model, shown below, incorrectly depicts the sugars and DNA bases attached to a phosphate "backbone." His theoretical structure was also not consistent with the data obtained from the partial hydrolysis of DNA, and the model was subsequently abandoned.

After much trial and error, in 1921, Levene proposed a structure for a strand of nucleic acid that was essentially correct. Levene's model, depicted below, correctly shows the DNA bases (thymine, adenine, cytosine, and guanine) bonded to the sugars of an alternating sugar-phosphate backbone. While beautifully elegant, Levene's DNA structure still possessed a serious flaw. In keeping with the tetranucleotide hypothesis, his model suggested equal proportions of the four DNA bases.

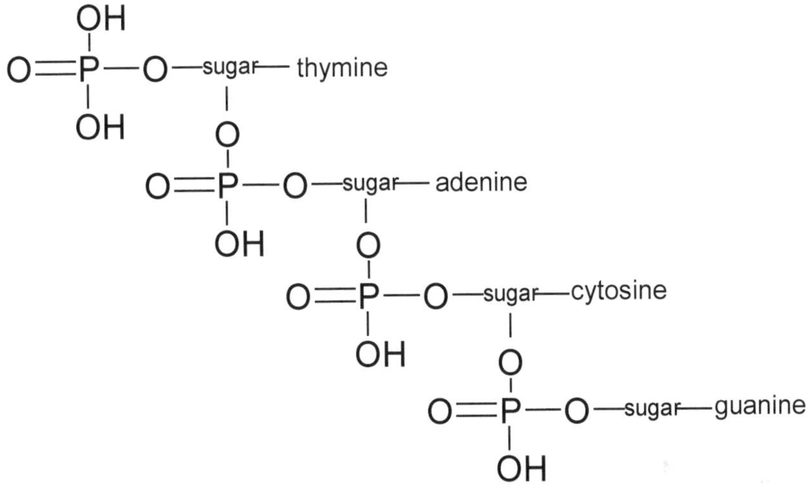

Several decades elapsed before scientists could develop the physical and chemical methods necessary to determine the three-dimensional shape of the DNA strands. One such method is x-ray diffraction, an indirect method of viewing a crystal structure, an orderly arrangements of atoms. When a beam of x-rays are shown on a crystal, some of the x-rays are diffracted, and consequently, are scattered in different directions. Some of the scattered x-rays reinforce each other and become more intense, while others cancel each other out. The x-rays then expose a photographic plate located behind the crystal, and by developing the film, one can observe the pattern created by these scattered rays. Using a number of mathematical formulas, the x-ray diffraction patterns can then be used to calculate various numerical attributes of the crystal structure.

The first x-ray diffraction photographs of DNA were obtained in 1938 by the research group of William Astbury, an English physicist and pioneer of the x-ray diffraction method. His group conducted x-ray studies of DNA fibers obtained from thymus glands. While Astbury's

hazy x-ray diffraction pictures showed that the DNA sample was actually a mixture of two molecular forms, which would later be called the
A and B forms, it more importantly showed that parts of the nucleic
acid were separated by a spacing of about 3.4 angstroms (1 angstrom is
equivalent to 1/10,000,000,000 of a meter) and that a repeat of some
sort occurred at about every 27 angstroms. Astbury and others correctly
surmised that the 3.4 angstroms value reflected the distance between
the nitrogen-containing bases, which were thought to occupy planes
perpendicular to the axis of the fiber. Astbury also believed, though incorrectly, that each sugar was in the same plane as its base. The x-ray
diffraction pattern also revealed that the DNA arrangement repeated
itself along the length of the fiber, which he mistakenly interpreted in
terms of the tetranucleotide hypothesis as the repetition of the bases
(i.e. TACG TACG TACG…). Among several models, Astbury considered a single-stranded, helical conformation for DNA, with the bases
located on the outside of the structure; however, such a structure could
not account for the 27 angstrom repeat, and it was subsequently dismissed.

In the late 1940s, Sven Furberg, a Norwegian crystallographer
working at Birkbeck College in England, proposed a different helical
model of DNA. Furberg's work with cytidine suggested that the sugar
and base of each nucleotide were oriented perpendicular to each other,
rather than co-planer as Astbury believed. He further proposed a single-
stranded, helical configuration for DNA, in which the bases were
stacked 3.4 angstroms apart and oriented perpendicular to the axis of
the helix. Unlike Astbury's model, Furberg favored a structure in which
the DNA bases were positioned on the inside of the helix. Furberg's
proposal was the first to account for the 27 angstrom repeat in terms of
a helix. Such a shape would explain the repetitive nature of DNA, for a
helix resembles a continuous circle when viewed on end.

Additional x-ray studies were conducted at Kings College in the
laboratory directed by English physicist Maurice Wilkins. Like other
scientists during World War II, Wilkins had used his talents for war-
related research including the atomic bomb project. Disturbed by the
application of physics to develop weapons of mass destruction, Wilkins
turned his attention to other areas of science. Wilkins was influenced
by Schrödinger's book *What is Life?* and decided to use his knowledge
of physics to study DNA. Wilkins had been given a DNA sample of
remarkably high quality from a Swiss scientist named Rudolf Signer.
At that time, samples had to be extracted directly from fresh tissues,

and the harsh methods used often broke the delicate DNA strands. Signer's refined method, however, resulted in long strands of intact DNA which gave excellent fibers for x-ray diffraction studies. Interestingly, Wilkins observed that the fibers could stretch reversibly between two forms. When dry, the DNA fibers were short and thick (A form). Though when sufficiently hydrated, they were long and thin (B form). Wilkins, suspecting that DNA was like a coiled spring, gave the DNA sample to Raymond Gosling, a doctoral student, to conduct preliminary x-ray studies. His initial photograph showed a blurry "X"-shaped pattern, an indication that DNA had a helical shape. Better data could not be obtained, however, because the x-ray diffraction instrument and techniques were insufficient. Fortunately for Wilkins, an expert in the x-ray diffraction technique had recently been hired – her name was Rosalind Franklin.

Franklin and Gosling, working together, provided the unequivocal experimental evidence that the DNA strands were helical. While Astbury and others had x-ray diffraction photographs of DNA, they were of low resolution and showed overlapping patterns of both the A and B forms. Franklin, on the other hand, developed a technique by which the x-ray diffraction experiments could be performed under high humidity – conditions in which Signer's DNA would occur only in the B form. In essence, Franklin created a method of separating the two DNA forms so they could be studied individually. The resulting x-ray diffraction patterns clarified the 3.4 and 27 angstrom values originally reported by Astbury's group. Her photographs indicated that the 3.4 angstrom spacing was a feature of the B form, while the 27 angstrom repeat described the A form of DNA. By 1951, Franklin correctly surmised that DNA was a helix composed of more than one strand, with the phosphates located on the outside. A year later, she and Gosling refined their technique and produced the now infamous x-ray diffraction photograph of the wet B form. From this photograph, they were able to infer that the B form helix had a spacing of 3.4 angstroms, a repeat of 34 angstroms, a diameter of 20 angstroms, and a slope of about 40 degrees. Franklin and Gosling also obtained excellent x-ray diffraction patterns of the crystalline A form, which turned out to be much more complex than those of the B form. From these data it became evident that the crystalline form of DNA had a monoclinic C2 space group. In other words, the DNA molecule appears the same when viewed from the top as it does when viewed from the bottom. Although Franklin misinterpreted the photographs of the A form, which she mistakenly believed was not helical, she clearly understood that the B form was a helix. While Franklin's approach was to collect experimental data first and then use it to derive the structure of DNA, two groups of scientists took the opposite approach. They built models of DNA first, and then waited to see how well the experimental data agreed with their proposals.

While Franklin and Gosling were busy collecting empirical data, two groups of scientists proposed that DNA was a helix, not of one strand, but of three. The three-stand model, or triple helix, was proposed by Nobel laureates James Watson and Francis Crick as well as by Nobel laureate Linus Pauling. Both groups believed that the orderly patterns exhibited by the x-ray diffraction studies were explained by the repetition of the sugar-phosphate backbone, rather than the irregularly shaped DNA bases. They envisioned three strands of nucleotides, with the sugar-phosphate backbone oriented towards the inside and the bases positioned towards the outside. Without knowledge of a mechanism by which the bases could be held and packed together inside of the helix, both groups suggested that the bases faced outward by default. However, the three-strand model could not explain the relatively high water content of crystalline DNA, and the model was not consistent with the values obtained from Franklin and Gosling's x-ray studies. Undeterred, Watson and Crick built a second model.

In 1953, Watson and Crick formulated their second helical model – this time consisting of two strands. Their first model had failed, in part, because they did not have access to Franklin and Gosling's data. Franklin was well on her way to elucidating the structure of DNA and did not see a need to collaborate. However, using Maurice Wilkins as a liaison as well as other sources, Watson and Crick were able to obtain the latest x-ray results without Franklin's knowledge. With such information, Watson and Crick built a two-strand helix with the DNA bases facing inward and the phosphates oriented outward, which would account for the high water content. Their double helix model had a 20 angstrom diameter, 3.4 angstrom base separation, 34 angstrom distance per helical turn, 10 base pairs per helical turn, with a 36 degree pitch. Furthermore, Crick correctly interpreted Franklin and Gosling's data on the crystalline A form to mean that the two DNA strands are oriented in opposite directions; one strands runs up while the other runs down. Watson and Crick's model of DNA fit Franklin and Gosling's data of the B form perfectly. The last unsolved problem was how to position the bases inside of such a helix.

   The problem Watson and Crick encountered with positioning the DNA bases was not new. It was a problem they had not addressed satisfactorily in their first model, and the reason why they had placed the bases on the outside of the helix. It was also not a problem with a trivial solution, for the issue had even perplexed Linus Pauling, considered by many to be the greatest physicist at that time. The answer was also not in Franklin and Gosling's x-ray diffraction data, which determined the shape of the sugar-phosphate backbone, but not of the DNA bases. According to Watson, "the real stumbling block was the bases," which presented a "frightful problem."[4]

   While Watson and Crick were mulling over this problem, a Catholic religious sister, named Sister Miriam Michael Stimson, OP, was experimenting with the very same DNA bases.

Marian (Miriam) Stimson and her younger sister, Alice Ruth.

Marian (Miriam) Stimson's high school graduation picture.

Chapter 5

A Bride of Christ

I will die to myself and my worldly past, to my personal history and my private identity. A new being, more holy and pure than the old, will rise from the ashes.

– Mary Gilligan Wong

Sister Miriam Michael was born Marian Emma Stimson on December 24, 1913 in Chicago, Illinois. She was the third of six children, the first girl, born to Mary Holland and Frank Stimson, English-Irish Catholic Americans whose relatives had immigrated to North America in the mid-1800s. Miriam was the daughter of a school teacher and a salesman. She spent most of her youth in Toledo, Ohio, but because her father was employed by a number of businesses, the family moved often. Her immediate family also included five siblings – two older brothers (Frank and Bill), a younger sister (Alice Ruth), and "the twins" (Charles and Catherine). Like her siblings, Miriam too could trace the origins of her reddish hair to her ancestors in Ireland.

Like many who shared Irish heritage, Mary and Frank Stimson were devout Roman Catholics and raised their children as such. The Stimson family attended Mass regularly and were active members of their parish – Mary Stimson even served as the organist. Consequently, the children attended parochial schools, where the classes were taught by religious sisters. Miriam's contact with the religious also included her own relatives, two of whom were Sisters of Mercy. As a child, Miriam would sometimes visit their convent, located behind the Chicago stockyards. While she was too young to contemplate following in

their footsteps, she did realize the hardships they faced, including toler-
ance of the foul stench of manure. During Miriam's youth, her family
too endured many difficult times.

Much of Miriam's childhood was filled with memories of family
illnesses. To Miriam, it seemed as if someone was always in need of
care. Miriam's older brother, for instance, was afflicted with polio
twice and needed constant attention. Meanwhile, Alice Ruth suffered
from rheumatic fever, a disease caused by a bacterial infection, which
severely damaged her heart. As a consequence of her poor health, Alice
Ruth had to attend a special school for physically handicapped children,
because her heart was not strong enough to allow her full physical activity. Miriam recalled that each morning "They [men from the school]
would come and lift her up [off the front porch] and carry her to the
school bus."[1] However, the Stimson family faced its greatest health cri-
sis after Mary Stimson gave birth to the twins. Following the birth,
Miriam's mother fell ill with eclampsia, complications due to high
blood pressure, and she lost her memory. "She [Mary] didn't know that
they [the twins] were her children. She called my brother 'little brother'
and my younger sister Catherine, 'little sister,'"[2] remembered Miriam.
Until her mother recovered, the care of the twins rested primarily on the
shoulders of Miriam, who was only ten years old at the time. Miriam
recalled that in the days before government-funded social service programs, she and her sister Alice Ruth would put the twins in a doll
carriage and leave them on the front porch with the family dog, a collie
named Lady, until the girls returned from school. "Lady was the protec-
tion for them,"[3] remembered Miriam. Her sister Catherine recalled that
during her mother's illness, Miriam "was doing everything. She seemed
to work the hardest and take the most responsibility. She would sew for
us. She would give us reading lessons. She had us doing the dishes."[4]
Eventually, Mary Stimson's memory returned; however, according to
Catherine, "she was never the same after that. She wasn't able to drive
any more and wasn't able to do all the things she had done before."[5]
Perhaps as a sign of the burdens placed upon her, Miriam, at age 14,
told her mother that she wanted to go (or rather escape) to China and
become a missionary. Her mother politely told her that she was too
young.

Miriam enjoyed attending school – it was a welcome reprieve from
her care-giving responsibilities. School provided Miriam with a sense
of freedom – to be all she could be intellectually. "I was interested in
exploring ideas,"[6] remembered Miriam, who was a model A student. In
Toledo, Miriam finished her primary education at the elementary

school of St. Agnes parish and then attended two years at Central Catholic High School (1928-1930). Miriam recalled that her eighth grade science teacher, Sister Agatha, initiated her interested in studying science. Sister Agatha was a young and energetic Sylvanian Franciscan, who at the time had just completed college. According to Miriam, Sister Agatha "inspired me with a desire to know more about God through investigations of the natural wonders of creation."[7]

Miriam's parents were supportive of their daughter's academic interests even though the climate of that time period did not support the education of women. In those days, women had few options other than marriage and motherhood. Since few women worked outside of the home, many felt that educating women beyond high school was unnecessary. Some even believed that an educated woman was unattractive to men and a college degree would diminish her chances of getting married. On the other hand, both Mary and Frank Stimson were well educated themselves – Mary Stimson was a certified teacher and a graduate of a Chicago normal school, while Frank received a business education from De LaSalle, a school sponsored by the Christian Brothers. Consequently, both parents understood the value of leading an educated life and encouraged this belief with their children. Miriam recalled that for an hour each day, "my mother, my two older brothers, and myself sat at the dining room table and practiced writing, forming letters, and all that kind of thing."[8]

When Miriam was a sophomore in high school, Mary Stimson decided that her daughter Alice Ruth should have a Catholic education, rather than the lay teachings she was receiving at the school for handicapped children. At that time, priests actively promulgated the belief that only the Church could provide divine truth and salvation. Out of concern for the soul of their daughter, the Stimsons looked for a Catholic institution that could educate and care for Alice Ruth. Mary then learned of St. Joseph Academy from her neighbor, and with an annual tuition of only about $300, the Stimsons found it the most affordable option. St. Joseph Academy was a Catholic boarding school for girls located in Adrian, Michigan, a small city located in southeastern Michigan approximately 35 miles northwest of Toledo and about 70 miles southwest of Detroit. The school had been established in 1896 by the Adrian Dominican Sisters.

The Adrian Dominican Sisters can trace their roots to Holy Cross convent in Regensburg (formerly known as Ratisbon), Germany. Holy Cross convent was home to a small group of contemplative-minded women who followed the rule for women in the Order of Preachers

(OP) established by St. Dominic. The sisters led prayerful, monastic lives under a vow of obedience, which included poverty and chastity. Other Dominicans adopted an apostolic life of itinerant mendicancy. In this manner, the Order combined their identity as itinerant preachers with the traditional monastic way of life.

Although the Adrian Dominican Sisters are sometimes referred to as "nuns," technically speaking they are religious sisters. Nuns live and work in monasteries, which have restricted access to the outside world. Nuns share a spiritual life of contemplative prayer and seldom venture outside of the cloister. Religious sisters, on the other hand, live in convents, which are more open to the secular world. Religious sisters live and pray within the convent, but may work outside in schools, hospitals, and other places of ministry.

The first Dominican sisters to arrive in Adrian, Michigan were sent from Holy Rosary convent in New York to staff two parish schools. Then in 1884, the Reverend Casimir Rohowski, C. PP. S., pastor of St. Joseph parish in Adrian, needed religious sisters to staff a hospital for railroad accident cases. The prioress of Holy Rosary convent responded to his request and sent six additional sisters to Adrian. They purchased an isolated farmhouse surrounded by cornfields in northeast Adrian and used it to establish St. Joseph Hospital. Although serving the needs of a few railroad accident cases, the Hospital became mostly a home for the afflicted, elderly, and orphaned.

By 1892, the Adrian Dominican Sisters faced "a state close to destitution."[9] The residents of St. Joseph Hospital were penniless, the sisters were earning pitifully small amounts for doing odd jobs at the local parishes, and the sisters were resorting to "begging tours." Earnings for 1891-2, for example, amounted to less than $700.[10] Furthermore, St. Joseph Hospital was no longer needed since railroad accidents had become rare, and when they did occur, the injured could easily be cared for by the newly built public hospitals. In addition to their financial problems, the sisters also experienced difficulties recruiting new members. Since few young women were joining the Adrian Dominicans and the elderly sisters were passing away, the congregation was slowly dying out. Mother Camilla Madden, the prioress, assessed the situation and made the decision to convert the hospital into a Catholic boarding school for girls. By doing so, the Adrian Dominicans could not only have a source of revenue, but also use the school as a means of communicating the religious life to young women.

In 1896, the sisters opened St. Joseph Academy and welcomed the first of many classes. At its opening, the school was described by the local press: "The Academy as it stands can accommodate from seventy-five to one hundred pupils. At the entrance on the first floor are four parlors, two on each side with half glass folding doors giving the rooms a cheerful and pleasing appearance. Five music rooms and a classroom are also on this floor. In three of the music rooms, ceilings of steel are handsomely decorated with paintings of Beethoven, Liszt, Mozart, Wagner, and other eminent composers. On the second floor we find a classroom, a large hall, and a beautiful and spacious chapel. The ceiling of the chapel is of steel and handsomely painted in delicate tints of blue, green, pink, and gold. …stained glass windows…large window of Dominican saints…statues of the Blessed Mother and St. Joseph…The third floor contains dormitories, wardrobes and bathrooms,…"[11] The sisters taught many courses including religion, English, music, drama, mathematics, French, and geography. The tuition was merely $2.50 per week, evidence of their dedication to making education affordable. Praise for the new school soon followed: "This Academy under the direction of the Sisters of St. Dominic is in one of the most pleasant localities of Adrian. The building is large and commodious and provided with all modern improvements. The grounds are extensive and well shaded, and afford ample advantages for outdoor amusements and healthful exercises. The institute offers every facility to aid young ladies in acquiring a solid, polite and Christian education, for while every attention is given to their advancement in the different sciences, no care is spared to train them to habits of virtue and to the refinements of good society. As moral training forms an essential element in education, the sisters spare no efforts to secure the development of mind and heart, considering themselves bound to act the part of parents for the children entrusted to their care."[12]

Although the academy began with only six pupils, the enrollment steadily increased each year. As more girls entered the academy and its graduates returned home, the reputation of the academy's outstanding academic stature spread. "No better institution for girls will be found in the State," declared Father Edward Joos, one of five priests who evaluated the school in 1899.[13] "The work of this Academy of learning has grown so rapidly that now it is in the front ranks of educational institutions of its character in the west," described a *Weekly Press* reporter.[14] The success of the Academy brought in much needed revenues. Furthermore, each year a handful of students would hear their calling in service to God and join the Adrian Dominican Sisters. Mother Camilla's dream of financial stability, rejuvenating the congregation with new members, and serving the spiritual and intellectual needs of young women in the greater community had come to fruition.

St. Joseph Academy was well established when Miriam and her sister Alice Ruth arrived in 1930. Miriam was sent to St. Joseph Academy, not only to finish her high school education, but primarily to look after her younger sister, a responsibility she took very seriously considering her sister's weak heart. Even though both girls were attending the same school, they were in different programs. Alice was in the elementary program, while Miriam was in high school. Consequently, Miriam did not see her sister for most of the day. In Madden Hall, Miriam shared a bedroom with three other students and a religious sister. This was the first time Miriam had spent a significant amount of time away from her parents. Miriam recalled her initial feelings of homesickness, while staring out her third floor dorm window and yearning for a glimpse of her parent's car. "I was very lonesome."[15]

On most days, the students of the Academy were too busy to think about their homesickness. Each morning, Miriam and the other girls awoke to the sound of a bell at 6:20 am, and then they had 40 minutes to wash, put on their blue uniforms, and make their beds before attending morning prayers and Mass. Afterward, the students ate breakfast in the refectory located in the North Building, which would later be replaced by the Weber Center, or in the dining room of Madden Hall. The students ate separately from the Adrian Dominican Sisters. Following breakfast, the girls attended their morning classes, which were taught by the religious sisters. Miriam recalled that they were very strict. They then recessed at 11:20 am and went for a morning walk. At noon, the bell rang again, and the students assembled for lunch. Afterward, they attended afternoon classes until 3:40 pm, when the girls received a bread and syrup snack before going outdoors for a second walk. Upon returning, the girls studied for an hour before supper, after which the students had another hour to study and complete their homework before attending evening prayers at 8 pm. An hour later, the girls retired for the night. The Academy did not have classes on Thursdays, but instead held classes on Saturdays.[16]

Life at the Academy was idyllic. In addition to taking classes, the girls performed chores in service to the school. They did a lot of cleaning, yard work, and sewing. Furthermore, the students enjoyed a lighthearted and simple social life. As recalled by one alumna, "We had dances, lots of music, silent movies, and plays. Recitals, far too long, and debates, often pointless, were integrated parts of academy life."[17]

Another former student noted, "We were at home, children of the household, and as such, we were not always well-behaved."[18] Some of the girls, for example, found amusement in "ditching" from their morning or afternoon walk. At other times, the girls would sneak out of bed around midnight and have a secret snack by candle light. On other occasions, the mischief occurred outdoors, as the girls would slip outside during the evening and raid the academy's garden for fruits and vegetables.

To Miriam, the Dominican sisters and her classmates had become a second family. Miriam was popular among her classmates, and she excelled in her studies. According to Sister Rose Louise Zimmer, OP, one of her high school classmates, "She was well liked. We were glad she was in our class. She knew all the answers."[19] "Marian was our class president," remembered Louise (Neckel) Godzina, another classmate.[20] In Miriam's senior year, her essay on Father William Doyle, entitled "A Present Day Saint," was published along with the writings of other students in the school's literary journal called "The Rosarian." In it, she extols the virtues of leading a saintly life, revealing her own deep admiration and respect for the religious life – a harbinger of the future.

A Present Day Saint

We often think a saint is just born 'that way' and it is easy for him to lead a holy life. To learn how erroneous this idea is we may look into the life of Fr. Wm. Doyle, SJ. True, Fr. Doyle is not a canonized saint, yet there is a reasonable possibility that he may be, for even in our own day this man practiced heroic virtue.

In the early part of the biography we are given an account of his youth and his entrance into the Jesuit Society. This is all very interesting but if we are really to profit by its reading we should closely follow his internal growth.

During his novitiate his spiritual life was very closely guided by the rule of the Society, but at this time he began the beautiful practice of saying a thousand aspirations a day; later in life he increased this number considerably. During his juniorate his spiritual life had less restraint but he seems to have continued in the same course as in his earlier religious life. Suddenly he was sent from his native Ireland to France to continue his studies. Here after his ordination he made his thirty day retreat. At this time he considered his past life very sluggish and he determined to become a saint, using as his motto: agree contra – to work against myself. We read also in the notebook, which he kept of his spiritual condition, that the things necessary for sainthood are a sensible love of God and the third degree of humility – that is a complete negation of all selfish and personal desires. At the conclusion of this retreat he also decided that God wanted him to offer himself for mission service. Fr. Doyle seems to have struggled with this decision but he yielded when he convinced himself that it was what God wanted, since he was restless and unhappy until he did give in. The Congo missions, he thought would afford an excellent opportunity to practice humility.

However Fr. Doyle's offer was not accepted, but instead he was given mission and retreat work in the British Isles. He was very successful in this field, working hard to obtain this success, the credit of which he gave to God. He is known to have spent hours in church during the night, that he might obtain the conversion of some soul. He would prostrate himself on the bare floor or he would pray with his arms outstretched until he thought they would break.

During this stage of his life he vowed to make a holy hour every day and he kept this vow under the most trying conditions.

Fr. Doyle served during the war as a chaplain to the Irish Fusiliers. One might think that this in itself would be sufficient penance to one of so sensitive a nature, but Fr. Doyle thought otherwise. Instead of diminishing his self-imposed penances he increased them. Among the others there is one very striking example: Fr. Doyle, in a place where food was very scarce, practiced abstinence of certain foods. He was always giving cake or candies which he received from home to his men. In return the soldiers idolized the joking, courageous priest and would do whatever he asked.

Fr. Doyle died a true soldier's death after a tedious life of living martyrdom at the front. Just as he wished

and prayed for a martyr's death, so he died for the Faith
– under fire, having performed his duties no matter the
cost – a true soldier in the army of Christ.

While all seemed well at St. Joseph Academy, in 1931, Miriam's
life unexpectedly took a more ominous turn. On one fateful November
evening, Miriam, Alice Ruth, and their friends were enjoying recreation
time in the Adrian room of Madden Hall. One of the students was playing the piano, while the other girls frolicked around the room. Miriam
recalled that Alice Ruth, with her weak heart, was dancing vigorously.
"She ought to ease up," Miriam remembered saying.[21] Later that evening, after the girls had retired for bed, one of Alice Ruth's roommates
awoke to a "gurgling" sound coming from Alice Ruth's bed. During the
night, Alice Ruth's heart had failed. The Dominican sisters were immediately called, and they transferred her to a private room. Miriam
was then wakened and told to telephone her parents that her sister had
died. Miriam was devastated, and even at the age of 88, got choked up
recalling the event. Miriam phoned, but could not bring herself to tell
her parents that Alice Ruth had passed away. Instead, she told her parents to come to Adrian and to bring their family friend, Mr. Blanchard,
a funeral director. Miriam recalled, "That was the closest that I could
come to saying that my sister had died... I couldn't say she died."[22] Not
only did Miriam feel utter despair, but she also felt that she had let her
parents down. She had been sent to Adrian to look after her younger
sister, and now she was dead. In Miriam's words, "I felt like I had
failed."[23]

The sudden loss of Alice Ruth was the nadir of Miriam's life.
Miriam and Alice Ruth had shared a special bond – a deep sense of
love, support, and loyalty that is uniquely shared by sisters alone. For
days, Miriam was prostrate with grief. Amidst feelings of remorse, failure, and loneliness, she slowly found consolation in the philosophical.
"We must sometimes feel pain, for a greater good."[24] Her faith un-
shaken, Miriam somehow managed to continue her studies, and she
gradually transformed her tears into knowledge.

In the spring of that academic year (1932) Miriam graduated
from St. Joseph Academy. The graduation ceremony was described
in the June 10 edition of the *Adrian Daily Telegram*, the town newspaper.

Thirty-one young women were presented with diplomas significant of a satisfactory completion of the

academic and college courses at St. Joseph's College
and Academy Thursday afternoon at the hands of the
Most Reverend Michael James Gallagher, D. D., bishop
of Detroit.

Miss Beatrice Cunningham, who with Miss Shober
graduated from the college this year, gave the address of
welcome to the bishop at the opening of the program after he had been escorted to the front of the large
auditorium during the processional of visiting priests.

A patriotic atmosphere was created in the program
when a chorus of students dressed in patriotic colors
beautifully sang "To Thee, O Country" after which the
graduates in appropriate costume presented a scene from
the life of George Washington. The scene was laid in the
Mount Vernon home of Washington and a feature of the
episode was a minuet by Washington, Martha Washington and their distinguished guests.

"The purpose of education is a supervised preparation for collective struggle against the forces of nature,"
declared the Rev. Austin G. Schmidt, SJ, dean of the
graduate school of Loyola University, who addressed
the graduates after the presentation of their diplomas.
"This is a physical world and its social structures have
been built up through the ages combating the forces of
hell which lead to the destruction of mortal souls. Youth
has idealism, hope, courage and charitableness while
middle age develops fear, satisfaction with what is and
hesitancy in attempting new projects. May you, this
class of St. Joseph's, bring to our land the spirit of youth
with its hope for the future and the opportunities which
you have been taught that God has placed in your
hands."

Father Schmidt further complimented the young
women of the school for receiving their education in a
smaller institution whereby the personal contact could
be made by the faculty and the students.

At the close of the program Bishop Gallagher extended his congratulations to the class and advised the
young women to go out into the world to lead, not to
step back and let someone else do it. "The need of lead-

ers in every department of life is great today, don't be bold but neither should you be too shy to take your rightful place in the world's activities," the Bishop said. Bishop Gallagher, who has just recovered from a several month illness, also thanked the Sisters, the student body and the audience for the prayers offered for him during his illness.

The orchestra, which is always an important factor in the entertainment of the hundreds of visitors at the academy at commencement time, was particularly good this year and added materially to the program of the afternoon. The first number was the processional, Brochton's "Commencement Grand March." Other numbers on the program included "Festival Overture" by Taylor and Percy Grainger's "Country Gardens." A recessional was also played by the orchestra.

Despite the empowering words of Reverend Schmidt and Bishop Gallagher, Miriam's future after high school appeared uncertain. She was an intelligent and beautiful young woman of strong faith, but what would she do next? Should she find a good Irish Catholic husband and become his wife and bear his children? Miriam had spent most of her youth as a child care giver, and she was not eager to jump back into that role. As would happen throughout her life, guidance came from the religious sisters. The Adrian Dominican Sisters also sponsored a woman's college, and the Sisters of St. Dominic offered to use the tuition previously credited to Alice Ruth at St. Joseph Academy to pay for Miriam's first year at St. Joseph College. Miriam graciously accepted the offer and began her college education in the fall of 1932.

St. Joseph College was a Catholic institution for women founded by the inspiration of Mother Camilla Madden. In order to educate the numerous young sisters, Mother Camilla established a college of liberal arts for the conferring of bachelor degrees and the preparing of teachers for certification. Several of the sisters, who had attended other universities, had earned their Master degrees and were qualified to form the first faculty. Encouraged by Bishop Gallagher, she initiated proceedings with the State of Michigan and obtained permission in 1919 to establish St. Joseph College.[25] "Though St. Joseph's College was officially opened in 1919, classes were taught in the academy building until the summer of 1922, when Sacred Heart

Hall was completed on the campus. In September of that year twenty-nine freshmen began their year's work, twenty-one sisters and eight young laywomen. Press notices hailed the new college building as the pride of Adrian. Referred to as a modern architectural masterpiece adding new beauty to the countryside, it was a splendid structure. In the center of its spacious marble hall stood a life-size statue of the Sacred Heart. In the words of Sister Benedicta Marie Ledwidge, who was then a young professed sister, the college was located 'far from noise and the smoke of the city, its windows looking out upon peaceful meadows and friendly groves, a scene which would delight a Wordsworth or Bryant in the spring time.' Sister continued, pointing out that the college 'lacked nothing that tends to develop an appreciation of the beautiful, and embodied the desire of Mother Camilla, its foundress, to provide an ideal environment for the study of truth and the pursuit of knowledge and culture'."[26]

In time, the failure to distinguish St. Joseph College from St. Joseph Academy became a serious matter, and in 1939, the name of the college was officially changed to Siena Heights College, in honor of St. Catherine of Siena (Italy). Saint Catherine of Siena is one of a limited number of canonized female Dominican saints. Many female saints, like Saint Agnes, are best known for choosing physical torture or death rather than surrendering their virginity.[27] It is by maintaining their chastity that these women saints are portrayed as making their greatest contribution to humanity. St. Catherine, on the other hand, is best remembered as a peacemaker who used her powers of persuasion to promote political resolutions to the schisms that divided the Catholic Church in the fourteenth century.[28] She represents a woman who empowers others through her message of active involvement, peace, and devotion. For these reasons, the private college for women was renamed after this female Dominican saint.

Student life at St. Joseph College was structured similarly to that of the Academy. As in her Academy days, Miriam resided in Madden Hall. "We were on the second floor of Madden Hall, in the west wing."[29] Each morning the young women would awake, get dressed, and then attend Mass. Afterwards, they would eat breakfast and then go to their morning classes. Following lunch, the students would attend their afternoon classes. At that time, all of the classes were taught by religious sisters and were held in a single building, Sacred Heart Hall. Miriam recalled, "The labs were [on] the fourth floor. The biology lab was down on the northeast corner, and the chemistry and physics lab was on the northwest corner... The library was on the third floor in the front of the building in a long room, and then there was a small room on the side that was theoretically the librarian's office, but she also kept rare books in there and things of that type. So we had those two rooms, and the card catalogue was out in the hall."[30] "I took biology and chemistry in my freshman year, German from Sister Regina Marie, and English [from Sister Jerome],"[31] Miriam remembered. Miriam's chemistry teacher was Sister Agnita Reuter, OP, whom Miriam credited as the teacher who got her interested specifically in chemistry. Sister Agnita was a friendly, yet strict, teacher. In those days, "If you had an assignment, you did the assignment. You didn't play around," noted Miriam.[32] As in her Academy days, Miriam excelled in her studies, especially in science. "It [science] came naturally to me," she would later attest.[33]

During her years at St. Joseph Academy and College, Miriam "grew to love, admire, and respect the Dominican sisters,"[34] and she in turn decided to become a Dominican. Having been taught in parochial schools about the deeds of numerous male saints, such as St. Francis and St. Thomas Aquinas, she saw in her Dominican teachers, strong *women* who were making a positive difference in the world, and who served as more tangible role models for her. One of the sisters, most likely, approached Miriam and asked her if she had ever considered a vocation as an Adrian Dominican Sister. She would then have been told that if she had even *thought* about it that she should give it a try for a year. "At least you'd know, then; it would be a lot better than living your whole life, always wondering if you had turned your back on a [religious] vocation."[35]

The news of Miriam's calling was met with family pride. While Mary and Frank Stimson would deeply miss seeing Miriam, they knew they were fulfilling their duty to the Church by allowing their daughter to follow in a holy cause. "When she decided to join the order, I'm sure they [Miriam's parents] felt that it was just fine," assured Miriam's sister, Catherine.[36] "Whatever any one of us wanted to study or do, our parents were totally supportive."[37] While Miriam's immediate family was supportive of her decision to join the Adrian Dominicans, her aunts, who were Sisters of Mercy, were not. They were upset that Miriam was not joining their congregation – an action they did not view as a matter of providence. Miriam remembered, "They were lovely, but they did not have the same outlook. They were very upset

when I became an Adrian Dominican. They called the Adrian Dominicans 'White-robed flappers'."[38]

Just prior to the start of Miriam's sophomore year at St. Joseph
College (1933), she joined the congregation as a "postulant," also
known today as a "candidate." "They used to call us postulants, from
the Latin "postulatio," which means to ask for things, because we
were asking to be admitted [to the congregation]," recalled Sister
Majella Gibson, OP, one of the more than twenty other young postulants that year.[39] The formal program of the candidacy was designed
to help the candidate discern her possible call to the Adrian Dominican Congregation by giving her six months to live this life. As a
postulant, Miriam lived in a community separate from the secular
students. Along with the other postulants, she wore a simple, black
dress "with long sleeves, with an apron on it, a cape, and a veil."[40]
She also resided in a dormitory room in Madden Hall and ate with
other postulants, separately from the sisters. Miriam also partici-
pated in liturgical services with the Adrian Dominican Sisters as
well as other social and ministerial activities. In addition to learning
about Dominican history and the mission of the Order, she continued
to attend classes at St. Joseph College.

The regimented schedules and activity-filled days left Miriam and
the other postulants very little if any free time. "With all of the other
things going on, we were supposed to carry on our full academic loads.
You don't have much time to study," remembered Miriam.[41] Consequently, Miriam utilized every possible moment to study. "She was
always studying, always studying," observed Sister Majella Gibson,
OP. "When we had to stand in line for something, she would concen-
trate on her little [flash] cards with her valences and things on it."[42]
Miriam remembered that even during recreation time, theoretically a
period for relaxation, they were to keep themselves busy. Mending
clothes was a popular "recreation" activity.

Several of the young postulants did not complete the year of candidacy. Not all the young women could adapt to the myriad of rules and
regulations that governed everything in their daily lives from what to
wear and eat to when to pray and sleep. As postulants, the women were
to leave their pasts behind and begin to adapt themselves to become
Adrian Dominican Sisters. Those who did not conform were strongly
reprimanded. Although the candidacy was billed as a "trial period,"
those who departed were stigmatized. Similar to Catholic divorcees at
that time, "ex-nuns" were often branded as failures. While some may

have wanted to leave, but continued because they feared the consequences, Miriam claimed she never entertained such thoughts.

After completing the first probationary period, Miriam and the other postulants entered the "novitiate" as "novices." The novitiate was a one-year period in which the women received instruction in religious life and engaged in reflection, community life, and personal and communal prayer. It was a time to expand one's contemplative spirit and to integrate more fully into Dominican life. Miriam and the others were formally initiated at a ceremony called "The Reception of the Habit," which was described in the August 9, 1934 edition of the *Adrian Daily Telegram*.

> With the impressive ceremony of reception and the solemn service of profession 22 postulants were received into the Catholic Order of the Sisters of St. Dominic and 39 novices took their first vows in Holy Rosary Chapel at St. Joseph's Academy this morning. The semi-annual service was attended by hundreds of relatives and friends of the young women and the chapel was filled for the ceremony. The altars in the sanctuary were beautifully decorated with flowers and lighted tapers.
>
> His Excellency, the Most Reverend John F. McNicholas, OP, S.T.M., Archbishop of Cincinnati, presided at the ceremony and preached the sermon in the absence of the Most Rev. Michael J. Gallagher, D. D., Bishop of Detroit, who is spending the summer in Rome.
>
> Pontifical mass was celebrated by the Most Rev. Edward F. Hoban, D. D., Bishop of Rockford, Ill., the Most Rev. Joseph C. Plagens, D. D., Auxiliary Bishop of Detroit, and the Most Rev. J. H. Albers, D. D., Auxiliary Bishop of Cincinnati. The Rev. Fr. James Cahalen, chaplain of the Academy, served as master of ceremonies and officers of the mass included the Rev. Fr. Donald Shields of Chicago and the Rev. Fr. James J. Marvin of Lainsburg, Mich., each having sisters who were received into the order, and the Rev. Fr. John C. McCauley of Chicago.
>
> The Rt. Rev. Monsignor John M. Doyle, D. D. of Detroit and a large number of priests from the dioceses and arch-dioceses of Cincinnati, Chicago, Detroit, Mar-

quette, Cleveland, and Toledo occupied places in the
sanctuary.

   After the services which concluded about noon an
informal reception was held for the Sisters and their
families and at 1 o'clock a luncheon was served to
nearly 400 persons.

   That morning, Miriam and the other young women removed the
black dresses they had worn during their year as postulants and adorned
themselves with white bridal dresses. On this day, Miriam would become a bride of Christ and faithfully devote herself to Jesus, who
would be her spouse. The wedding dress worn by Miriam was part of
the congregation's collection, donated by the families of the sisters and
worn by other young sisters on this occasion in years past. At the beginning of the reception of the habit, Miriam joined the procession of
radiant brides as they solemnly walked down the center aisle of Holy
Rosary chapel, well known for its architectural splendor – its vaulted
ceilings, stained-glass windows, and ornate altar piece. As the brides
filed past each wooden pew and the women that would become their
spiritual sisters, Miriam's eyes may have diverted momentarily to her
parents, seated in the audience. Despite the nuptial imagery, this wedding was unlike any other. Instead of dreams of wedding gifts and of a
honeymoon, Miriam and the other Christ brides were about to join their
crucified bridegroom – like Christ, they too were to die. On this occasion, they would surrender their lives and relinquish any men they
might have loved and any children they might have had.[43] Like Christ,
the women would sacrifice their secular lives for the salvation of the
world.

   During Mass, the brides approached the altar, and the Archbishop
inquired, "My dear sisters, what do you ask?" In unison, the brides replied, "I ask to be permitted the habit of the Dominican Order." After
completion of the blessing of the habits, each bride stepped forward,
received a neatly folded habit, and then silently departed the chapel in
order to change. Miriam and the others then donned the habit for the
first time and completed the symbolic immolation.

   The habit of the Adrian Dominican Sisters was representative of the
style of clothing worn by the German peasantry in the late medieval
period. In contrast to the black robes commonly worn by other Catholic
nuns, the Sisters of St. Dominic wore white. Their habit consisted of a
white, long-sleeved, floor-length tunic secured around the waist by a
thin, black belt from which hung a long, 15-decade rosary. The front of

the dress was covered by a white, rectangular, ankle-length scapular, while a white collar draped over the shoulders. A white wimple was worn to cover the neck, ears, and forehead, while a black, white-lined, veil was worn over the head and draped over the back shoulders. When traveling outdoors, a black cloak was worn.

An unusual event occurred during the reception of the habit. After donning the habit for the first time, the sisters formed a line and approached the Archbishop, one at a time in alphabetical order, to receive their new name. According to Sister Majella Gibson, OP, "The concept of entering a religious order implied that you gave up everything that by which you were known before, and so it involved changing your name, and otherwise changing your identity... Many of the sisters would take the name of the saint that their parents were named for."[44] Sister described the following events: "Four people led the line. The first four of us carried a candle and a cross."[45] However, the priest assisting the Archbishop, calling out the names and handing the cards to the Archbishop, accidentally skipped a name, so most of the candidates received names other than the names they had requested.[46] Sister continued, "The person in back of me received the wrong name. She was sort of a meek person, so she didn't make a very loud protest. The Priest had a list, and she tried to tell him that wasn't her name, but he just went on."[47] Miriam was supposed to have received the name of Mary Giles, but because of the mix-up, the sister in front of Miriam received that name. So when the Archbishop reached Miriam, the last woman in line, he had run out of names. The Archbishop then quickly thought of another name and pronounced that she would be called Sister Michael, in honor of Bishop Michael Gallagher of the Detroit diocese. However, the congregation already had a Sister Michael, so Miriam became known as Sister Miriam Michael. Later, when it became customary to use the family name, she would go by the name of Sister Miriam Michael Stimson.

Over the next three years, Miriam would integrate more fully into the community of the Adrian Dominican Sisters. After receiving the habit, Miriam began the required canonical year, during which she and the other novices focused on the development of a personal, contemplative, prayerful life and prepared themselves for a vowed commitment as Adrian Dominicans.[48] After completing the canonical novitiate, Miriam then professed her vows to the Prioress on August 13, 1935 – vows she would renew annually until professing perpetual vows five years later. Unlike other congregations which take separate vows of celibacy, poverty, and obedience, Dominicans only make one vow – obedience.

Miriam explained, "The Dominicans feel that everything comes under the will, and obedience is where you turn over your will."[49] Upon first profession, Miriam officially became a member of the Adrian Dominican Sisters and made the Congregation's mission a part of her life. She then received the black veil, an indication that one was a professed religious of the congregation, to replace the white veil she had worn as a novice. During this time, Miriam also completed her studies in chemistry, and in the spring of 1936, she received her bachelor's of science degree from St. Joseph College. Little did Miriam know that she would soon be beginning her long scientific career studying the bases of DNA.

Sister Miriam Michael Stimson, OP, and Dr. Elton Cook
at the Institutum Divi Thomae.

The 1939 graduates of the Institutum Divi Thomae.
Sister Miriam Michael Stimson, OP is the second woman from the right
standing in the top row. Her friend, Sister Mary Jane Hart, OP is the
second woman from the left sitting in the bottom row.

Chapter 6

An Act of Worship

Science is not a religion, but scientific research
is a principal act of religion.
— Sir William Bragg

ollowing their profession of vows, the Adrian Dominican Sisters were usually "sent out on mission," away from the Mother-house, to teach in parish schools or to perform other kinds of missionary or service work. However, this tradition gradually changed in the 1930s, when Mother Mary Gerald Barry, OP, was elected Prioress (Mother General). Soon after assuming the office, Mother Gerald began to vigorously promote the higher education of the sisters. Most of the sisters received their undergraduate education at St. Joseph College, later re-named Siena Heights College. Afterwards, many were sent to other institutions to further their studies and to earn Master and Doctoral degrees. Upon completion of their advanced degrees, the sisters then joined the teaching faculties of the congregation-sponsored colleges, like St. Joseph College.[1] "We were blessed with having Mother Gerald, because she really valued education," recalled Miriam.[2] So instead of leaving the Motherhouse after taking her first vows in the fall of 1935, Miriam was retained in Adrian to complete her studies.

In the spring of 1936, Miriam, wearing a graduation gown over her habit, joined the procession of students in line to receive her diploma. Once again, Miriam walked by the beaming faces of her parents, who were in attendance as they had during Miriam's graduation from St. Joseph Academy and reception of the habit. When Miriam's turn finally came, she stepped forward and received her diploma from none

other than the Reverend Michael Gallagher, the Bishop after whom
Miriam was named. One can only imagine the sense of providence that
she may have felt at that moment. Other events of that day, as well as
the commencement address, were described in the June 4 and 5 editions
of the *Adrian Daily Telegram.*

> The ceremonies accompanying the commencement
> exercises at St. Joseph's College and Academy today
> opened this morning with a Community Processional,
> the entire group of Sisters of St. Dominic and the stu-
> dents of the institution marching to Holy Rosary Chapel
> where solemn high Mass was conducted. Led by the
> Rev. Mother M. Gerald, president of the college and
> Mother Superior of the Order of the sisters of St. Domi-
> nic, and Sister Benedicta Marie, dean of the college,
> followed by 250 Sisters of the community and the stu-
> dents, the class of 1936 was thus honored.
>
> The Most Reverend Michael J. Gallagher, D. D.,
> Bishop of Detroit, was present today to confer the hon-
> ors at the exercises this afternoon and others of note who
> were being entertained by the Sisters today were the
> Very Rev. Henry Ignatius Smith, OP, Ph.D., LL.D., pro-
> fessor of philosophy of The Catholic University of
> America at Washington, D. C., and the Rev. Mother
> Mary de Lourdes, Mother General of Mt. St. Mary's
> Academy at Newburgh, N. Y., and her companion Sister
> M. Anita.
>
> The program opened this afternoon with a grand
> march by the orchestra which played as Bishop Galla-
> gher and the visiting priests entered the auditorium
> where the exercises were held. They were greeted by the
> "Hallelujah Chorus" from "The Messiah" after which
> the orchestra played "Cosi Fan Tutti" (Mozart).
>
> After the march of the graduates, the address of wel-
> come was given by Miss Elizabeth K. Burns. After the
> conferring of the honors by the Rev. Bishop Gallagher,
> the address of the afternoon to the graduates was given
> by the Very Rev. Henry Ignatius Smith.
>
> In a commencement address delivered at St. Jo-
> seph's College and Academy yesterday afternoon the
> Very Rev. Henry I. Smith professor of philosophy at the

Catholic University of America in Washington, D. C.,
commended the Dominican Sisters on the excellence of
their record of teaching. He pointed out that behind the
Adrian School stands the traditions of 700 years of
higher learning and higher education.

He discussed the principles behind Catholic education and the high efficiency that education has attained,
pointing out to the graduates that future critics will
judge the value of that education by what they produce
in practical American life.

His address in part follows:

"The graduates of St. Joseph's College are particularly fortunate in having the training given by the
Dominican Sisters. Behind this school stands the traditions of 700 years of high learning and higher education.
Since the 13th century the Dominican Order has conse-
crated itself to advanced research and scientific
teaching. During the last decade no branch of the
Dominican family has shown such strides in progressive
improvement of educational methods as the Sisters of
your college.

Behind Catholic education are four principles. Without a knowledge of these it is impossible to understand
our whole educational progress. They have inspired
Catholic educational activity from the kindergarten to
advanced university research for more than 1,500 years.
These principles are first, that the Author of this world
has a right to be known by every creature born into society. He has a right to be known in His most intimate
contacts with the sciences and with life.

The second principle behind Catholic educational
system is the right of every child born into society to
know its nature, its origin, and its destiny. Since the
child has a right to such knowledge, society is under the
unescapable obligation of providing it.

The third principle inspiring Catholic education is
the right of society and the nation in particular to receive
from our schools future citizens trained and disciplined
in social duties and national virtues. No educational sys-

tem has a right to exist within a republic unless it does train for citizenship. We of Catholic belief are firmly convinced that both the nation and civilization profit from training which offers through religion not only an idea of civic virtue but unescapable reasons for practicing it.

The fourth principle behind the Catholic educational system is our conviction that faith must be kept with education itself. The school ought to be permitted to develop the whole child, to develop all of character, not merely the physical or the mental phases of them. This means that education should be given the opportunity to develop the powers of the soul. And until it is given such opportunity to develop the soul, education in the state and by the state should not be blamed for its failure to produce character.

These principles upon which our educational work is built have been defended and realized in the United States at the cost of great sacrifice. This sacrifice is represented in the heroic struggles of our Catholic teachers, underpaid and overworked, and yet satisfied to consecrate their lives for the fulfillment of these sublime purposes. This sacrifice is also represented in the generous self-denial of your parents who have shouldered the responsibilities of taxation for the state education and who add the burden of supporting our church schools at the same time.

We are firmly convinced that our sacrifices in educational work during the last 150 years in this country have not been made in vain. We think that the general verdict of the American people is in agreement with our convictions. We know that educators and statesmen marvel at the high efficiency which Catholic education has reached under the tremendous handicaps with which it has been burdened. We are certain that you graduates go out into the world with equipment for which you need offer no apology, but it is well for you to remember that the verdict of future critics will be rendered on what you produce in practical American life.

You will command respect for our education and for yourself if you bring wherever you go in teaching, in the business world, in social life, those qualities which America needs most and which so few of our citizens are prepared to contribute. I speak of those primitive and rugged American virtues by which the republic was sent forth on its glorious course… I speak of obedience to law and reverence for authority. I speak of purity, individual, domestic, and public. I speak of justice and fair play to both the rich and the poor. I speak of godliness, no matter what be the merits of social planning and social reconstruction, they will be vague and vain dreams unless they can be built upon these virtues. In practicing them you serve not only your church, not only this institution to which you owe loyalty, but you serve your God. They are contributions which we pray that you will make wherever you live and whatever else you do."

Following the commencement, Miriam received her first mission assignment, an occasion often filled with anxiety. The uncertainty felt on such a day was recounted by a former religious sister. "One might be sent to teach grade school or high school, to a rich parish or a poor one. The Superior may be understanding and supportive, or she could rule the house (and you) with an iron hand. If you teach in a parish grade school, the pastor may be a benevolent father or an eccentric tyrant. The sisters in the house may be fun and young-at-heart or a bunch of frustrated sour old ladies. Most important of all, you may be sent to a location within a few minutes of where your families lives, or you may be sent thousands of miles away and not see them for years."[3] During her meeting with Mother Gerald, Miriam was instructed to go to Cincinnati, Ohio and earn a Masters of Science degree from the Institutum Divi Thomae, a newly created Catholic graduate school. Miriam was pleased with her assignment. Not only could she pursue her passion for learning, but she would also be located in a city where her parents only needed to drive a couple of hours in order to visit her – Miriam, herself, could not drive, because the veil of her habit blocked her peripheral vision. Mother Gerald would later be the first thanked by Miriam in her completed dissertation. After renewing her vows, Miriam prepared herself for her first assignment.

Whenever the Adrian Dominican Sisters left the Motherhouse, they always traveled in pairs. The sisters journeyed in pairs not only to make it easier to navigate unfamiliar avenues, but also because they were forbidden ever to be alone with a man. On this journey, Miriam was accompanied by Sister Mary Jane Hart, OP, who would attend the Institutum Divi Thomae as well. Sister Mary Jane, like Miriam, was born in Chicago to parents of Irish descent. She had attended high school at St. Xavier Academy and had been taught by the Mercy Sisters, the same order of religious sisters that Miriam's aunts belonged to, before she had decided to become a Dominican. Though unlike Miriam, Sister Mary Jane had been sent out on mission immediately after her profession of first vows. She then spent six years teaching math and science at two different parish schools before returning to St. Joseph College to complete her Bachelor's of Science degree.[4] Miriam and Mary Jane were the first Adrian Dominican Sisters sent to Cincinnati to study.

In Cincinnati, Miriam and Sister Mary Jane resided in a Dominican convent located in a slum district of the city. There, she learned first-hand what was meant by overcrowded housing, lack of recreational facilities, and sickness due to poor sanitation and undernourishment.[5] In sharp contrast to the idyllic environment she had experienced in Adrian, in Cincinnati Miriam had to put up with the constant noise and even shootings.[6] Each morning, Miriam and Sister Mary Jane attended Mass at a local church and then received a ride to the Institutum Divi Thomae. When they arrived at the Institutum, they were joined by a coterie of forty- and fifty-year-old religious sisters from other congregations, including the Sisters of Charity of St. Vincent de Paul, the Religious Sisters of Mercy, the Sisters of the Most Precious Blood, and the Sinsinawa Dominican Sisters.[7] Among these "nun-students," Miriam, at age 23, was by far the youngest.

The Institutum Divi Thomae, Graduate School of Scientific Education and Research, was established in 1935 by the Most Reverend John T. McNicholas OP, then Archbishop of Cincinnati, and Dr. George Speri Sperti. During the Depression years, the Archbishop saw communism as a growing threat for Catholicism. "Some day the atheistic communists are going to be a force in the world, and science will be their big tool," the Archbishop told Sperti. "We must keep pace with them."[8] With a desire of advancing science within the framework of Catholicism, the Reverend John McNicholas formed a fourth unit of the Athenaeum of Ohio, a graduate school devoted to scientific research and education. The graduate school was named the Institutum Divi Thomae after St. Thomas Aquinas, a Dominican patron saint of students and theologians.

Dr. George Sperti, who was then only 35 years old, was selected as the director of the Institutum Divi Thomae. Before joining the Institutum, he had been the director of the University of Cincinnati's Basic Research Laboratory and had earned the reputation as a brilliant, young biophysicist, having been awarded patents on fluorescent lighting, vitamin preparation, food preservation, meat tenderizing, and food irradiation. Although highly successful, Sperti's work at the University of Cincinnati was limited by a lack of research funds and equipment. The chance to direct the Institutum Divi Thomae, on the other hand, gave him the scope and opportunities he was looking for: a modern laboratory in Cincinnati, a small but carefully picked staff of research scientists, and a major project – cancer research.[9]

To foster a program of cancer study, Sperti organized a research foundation under the leadership of Charles F. Williams, who was then president of Western-Southern Life Insurance and a major benefactor of the Institutum.[10] The purpose of the foundation was to hold the rights to marketable patents developed out of the research at the Institutum and license them to others for manufacture.[11] The foundation played a crucial role for the Institutum, since the monies collected from the licensing of these rights comprised about half of the income for the school – the other half coming from private contributions.[12] These revenues were used not only to fund the cancer research and salaries of the scientists, but also provided free tuition for Miriam and the other nun-students.[13]

In planning the Institutum Divi Thomae, Sperti could have created another undergraduate college, but instead decided on a graduate school. Even in the 1930s, Sperti realized that higher education would become as commonplace as secondary education, and one day a Bachelor's degree might be as ordinary as a high school diploma.[14] Consequently, he saw a need for a school for post-graduate work and research, rather than another school for undergraduate education. Two types of degrees were granted by the Institutum Divi Thomae. The first, that of a Master of Science, was conferred to students who had completed three years of research and had contributed to the publication of at least three scientific papers. The second degree, the Ph.D., was awarded to graduates who had made significant contributions to their field of study after leaving the Institutum.[15]

After establishing the Institutum Divi Thomae, Sperti created a network of affiliated hospitals and colleges. Sperti formed ties with physicians at hospitals in Illinois, Ohio, and New York, in order to conduct clinical trials on potential drugs derived from the research conducted at the Institutum. Furthermore, after receiving their Master's of Science degrees, the Sisters returned to their home colleges and established branch units of the Institutum Divi Thomae. Units were formed at Rosary College, Marymount College, Barry College, and Siena Heights College, where the Sisters were expected to continue their cancer-related research in association with the Institutum.[16]

When Miriam was a student, the Institutum Divi Thomae was housed in a wing of St. Gregory's Minor Seminary. The research wing consisted of 4 to 6 laboratories constructed on a wooden floor that covered an old bowling alley.[17] Later, the Institutum expanded to include the music room.[18] In 1940, the Institutum moved to Walnut Hills and occupied several buildings on Madison Road that had been donated to the Church.[19] While the focus of the research was always on cancer, the work resulted in a number of practical applications, mostly in the enrichment of various skin care products such as soaps, lotions, and ointments. In addition to the buildings in Cincinnati, the Institutum acquired a former gambling club in Palm Beach, which was used to establish a Florida branch. During the Second World War, researchers there discovered a new source of agar to replace the supply that had been cut off due to the conflict. In all, the discoveries made at the two branches resulted in 127 patents, the royalties of which were used to fund the Institutum and Sperti's numerous entrepreneurial interests.[20]

Sperti was a devout Catholic, and with the approval of Archbishop McNicholas, he established the Institutum as a school of science based on the philosophical riches of Christianity. Sperti was also a creationist, who viewed God as a "Divine Intelligence." He believed that the physical construction of the universe and its operation, as revealed through science, gave ample and absolute evidence of a Creator. Consistent with such beliefs, Sperti viewed the process of science as "a chain of logical discovery" dependent on an orderly and predictable universe, which would turn chaotic in the absence of God. Sperti saw God in the heart of nature and rejected atheism as being, above all, unscientific.[21]

The general philosophy of the Institutum, as conceived by Sperti and Archbishop McNicholas, was to demonstrate that there is no conflict between science and religion, but rather that science strengthens religion and the belief in a personal God.[22] To achieve this, Institutum scientists were to investigate the basic laws of nature and to interpret the findings, with the assistance of priests, in accordance to the tenets of Catholicism. To promote this philosophy, the Institutum used a number of slogans. "The Church has no fear of even the most daring progress of science, if only it be true science" – Pope Pius XI, and "Toward a clearer understanding of the basic laws of nature for the betterment of mankind." were used in various publications.[23] Furthermore, the Institutum created an insignia, which combined both scientific and religious symbols. The insignia consisted of a young woman, dressed in white, standing above an enlarged mortar and pestle in front of a cross, looking upwards at "the light of Christian philosophy" held between her outstretched arms.[24] According to Miriam, the Archbishop, Sperti, and the Institutum exemplified their view of combining science with religion.[25]

To help bridge the gap between science and religion, the Institutum Divi Thomae was staffed with scientists, who were philosophically-inclined, and priests, who also served as philosophers.[26] The faculty of the Institutum Divi Thomae was divided into two groups – research professors, who conducted experiments at the Institutum throughout the year, and lecturing professors who taught specialized subjects. When Miriam was attending the Institutum, the research faculty consisted of a diverse mix of scientists who specialized in a wide range of disciplines including physiology, biochemistry, physics, mathematics, bacteriology, and plant physiology.[27] Although the Institutum staff possessed markedly different backgrounds, in one way or another, each faculty member tied his science to God.[28]

The emphasis of the Institutum on combining science with faith most likely influenced the way Miriam viewed her research. Historically, science and the Catholic Church have had an antagonistic relationship; however, Miriam found that science and religion can work side by side.[29] Throughout her life, Miriam blended her faith with her view of the natural world. For instance, in 1938 she wrote, "The fundamental mode of operation of light is outside the limitations which man might set by his formulations – it is objective. As in the solar spectrum, so too, the fundamental laws of morality are outside the influence of time and custom; they are objective."[30] Furthermore, over sixty years later, Miriam would attest, "My Catholic identity is how I see the world, and the world is dependent on God."[31] Miriam viewed God not only as the creator of all things, but also as being represented through the physical and chemical laws that governed the natural world. By studying these chemical phenomena, Miriam believed she could strengthen her relationship with God. Similar to the way that one becomes familiar with artists through their art or gets acquainted with authors through their writings, Miriam got to know her Creator by studying creation.[32] In this manner, Miriam viewed her scientific pursuit of truth as an act of worship.[33]

To further encourage the synthesis of science with religion, the methods of teaching at the Institutum Divi Thomae were based on the traditions of the medieval seminaries and universities – a few students were assigned to work under a mentor, who gave them individualized instruction. While Miriam and the other students attended a limited number of formal classes, the greatest part of their work and learning occurred in the laboratory, where their instruction revolved around the field of science in which they were specializing. By this method, each student was expected to obtain a thorough understanding of the field in which she was studying, but at the same time, she was expected to acquire a comprehensive knowledge of all the sciences. Although each student was assigned her own project to work on, the research was part of a collective endeavor that was guided by the professor to whom the student had been assigned. By this approach, each student could then benefit from the viewpoints and results of the other students working under the same professor. Furthermore, since all of the research professors worked on some aspect of cancer, the students could also learn and gain from the work conducted in different, yet related, areas of science.[34]

At the Institutum, Miriam received most of her instructions from Dr. John Loofbourow, under whom she was assigned to work. Miriam recalled, "We had some general kinds of courses, but the idea [was that] there would be groups. For instance, in my group, Loofbourow was the man in charge, and there were four or five of us who worked in that group. So it was like a small research group. Instruction was done around the table, with Loofbourow at the head of the table. We had places around the table, the library books were around us, and we got instruction in that kind of way. It was very different from the formal thing at the big universities. You began to learn the things you needed to know. You got some physics, you got some math, and you got some genetics. You learned what you needed to do the research that you were doing."[35] In addition to science, "one of the priests from the major seminary always came to teach philosophy," added Miriam.[36]

In the laboratory, Miriam conducted her research under the directions of Dr. John Loofbourow and Dr. Elton Cook. Loofbourow was a biophysicist, who had met George Sperti at the University of Cincinnati before joining the Institutum Divi Thomae. As a scientist, Miriam described him as "a good chemist," and "the kinds of questions he would ask me made me look at things from another perspective."[37] As a person, Miriam remembered him as being "very open," "encouraging," and "congenial."[38] Likewise, Miriam had fond memories of Elton Cook. Prior to joining the Institutum, Cook had been a research biochemist at the William S. Merrell Company. According to Miriam, "He brought another perspective – to look at the practical aspects of chemical research."[39] Miriam recalled that both Loofbourow, whose wife was a physician, and Cook were very supportive of women in science. While Miriam credited Loofbourow for initiating her interest in spectroscopy, she acknowledged Cook as the person who got her interested in organic compounds, like DNA.

Having male instructors was a new experience for Miriam. Apart from one pedant at St. Joseph College, in all her years of schooling, she had been taught only by religious sisters. After living six years in Adrian with limited contact with men, other than aged priests, she was now conversing on a daily basis with persons of the opposite sex. Miriam's contact with men was primarily with Loofbourow and Cook, although there were a number of other male professors too. Loofbourow and Cook were highly intelligent and successful scientists. Both were married, and Cook had children. Miriam also had conversations with the charming, handsome, and unmarried director of the Institutum, Dr. Sperti, who was only a few years older than Miriam.[40] At that time, "Dr. Sperti was young and a hot property," described one reporter.[41] Surrounded by maleness like she had never experienced before, Miriam developed a deep admiration for these men and formed life-long friendships. On occasion, the wives of Loofbourow and Cook would visit the Institutum. Visions of a woman holding her spouse and child may have entered Miriam's thoughts, as well as imagining what it would have been like to be a wife and mother.

When Miriam joined the Institutum Divi Thomae, Sperti, Loofbourow, Cook, and the other scientists were coordinately working on cancer research – conducting experiments involving cell damage and cell division. They were exploring the normal but enigmatic phenomenon that when a living tissue is injured, the cells surrounding the wound break out of their quiescent state and begin to reproduce at a furious pace. Only when the destroyed tissue has been replaced and the wound healed do the cells return to their normal state. But what instructs a cell to divide or not to divide? For more than half a century, scientists have speculated that the answer lies in the existence of chemical "wound hormones." Under the premise that cancer is the result of uncontrolled cell division, Sperti and the other scientists at the Institutum decided that the best approach to understanding cancer was to isolate and characterize the wound hormones.[42]

The first step in this process was to develop a method of damaging cells without destroying them. But what kind of cells should they use? For years they tried cells from different organisms, such as fish, lizards, chicken, and yeast, and also from different tissues, like liver, kidney, and embryo. Eventually, they settled on conducting most of their experiments with yeast, in part because the yeast cells were similar to human cells and were easy to grow in the laboratory. To injure the yeast cells, the researchers at the Institutum used ultraviolet rays from the lamps that Sperti had previously patented. After many unsuccessful trials to determine the quantity of cells to use, the UV intensity, and the time of exposure, they eventually established the conditions upon which they could consistently injure the yeast cells without completely destroying them.[43]

To determine whether the injured yeast cells were releasing wound hormones into the liquid medium in which they were grown, the researchers removed the yeast cells by filtration and then tested the remaining cell-free filtrate. If the filtrate contained such a chemical substance, then it should be able to stimulate the growth of normal yeast cells. When tested, that effect is exactly what the researchers found. Furthermore, neither the filtered cells nor the filtrate obtained from untreated cells possessed any growth-promoting activity.[44] Additional experiments were performed to determine whether this phenomenon was unique to yeast and UV radiation. It was not. Like the UV irradiation experiments, similar results were observed using the cell-free filtrates obtained from yeast cells damaged with x-rays, heat, and mechanical injury.[45] In addition to yeast, comparable results were obtained using irradiated liver, kidney, and embryo tissues.[46] After many years toiling away in the laboratory, the researchers had finally accomplished their goal – discovering some of the first cellular factors involved in cell division. Realizing the potential implications of their discovery relative to cancer, the growth-promoting, intercellular, wound hormones were termed "biodynes," derived from the Greek words for "life-force."[47]

To study the relationship between the biodynes and cancer, the Institutum researchers exposed the yeast cells to chemicals that had long been suspected as carcinogens. Although UV radiation is a cancer-inducing agent, at that time the harmful effects of excessive exposure to this form of radiation was neither realized nor was it known to be implicated in skin cancer. As a consequence, the researchers believed that experiments conducted with cancer-causing "chemical irritants," rather than UV radiation, were a better test of the link between the biodynes and cancer. When these experiments were performed, the researchers discovered that, like UV light, the chemical irritants damaged the yeast cells and the filtrates were found to contain growth-promoting activity. These results were interpreted in light of the chronic irritation hypothesis of cancer. Sperti proposed the idea that the chemical irritants "have the power to injure large numbers of cells, and to keep them injured over a long period of time, resulting in the secretion of a large and continuous quantity of growth-factor [biodynes]."[48] Sperti believed that the biodynes, rather than the chemical irritants, were the direct cause of the uncontrolled cell division seen in cancerous tissues.

The investigators then turned their attention to the biodynes themselves. They discovered that the cell-free filtrate contained a mixture of different substances, which were claimed to affect cells in different ways. One type of biodyne was found to induce cell division and was believed to play a role in wound healing. Another type was observed to stimulate cellular respiration, the uptake of oxygen, while a third biodyne was found to accelerate the metabolism of sugar.[49] These different types of biodynes were thought to function collectively as the roots of abnormal cell division – not only by initiating cell division, but also by accelerating the growth of cells by speeding up the consumption of oxygen and the utilization of energy-containing sugars.

After joining the Institutum, Miriam was assigned the project of studying the chemical nature of the biodynes using ultraviolet spectroscopy. UV spectroscopy is a method of characterizing a chemical compound based on its absorption of ultraviolet light, the region lying just beyond the violet end of the visible light spectrum. "When you go to the cafeteria, you are engaging in a basic sort of spectroscopy," Miriam explained. "Your eye studies the interactions between visible light and the chemical composition of, say milk and water. From the differences in the way the two absorb and reflect light, you can tell which is milk and which is water."[50] "In the same way we measure chemicals under infra-red or ultra-violet light. We learn by plotting graphs of the wavelengths what the chemical properties are."[51] When Miriam was a graduate student, spectroscopy was in its infancy. In contrast to the continuous spectra that can be easily obtained and plotted by computerized instruments today, Miriam had to photograph the UV absorption spectrum and then meticulously draw each graph, point by point, by hand. Although laborious, Miriam recalled, "I loved it [spectroscopy] right from the start. However, I will confess it took me a month before I could correctly pronounce it. My friends would razz me about it and say I was studying specks."[52]

The first step in her project was to compare and contrast the UV spectra of the filtrate of irradiated yeast cells to that of non-irradiated cells. To prepare the crude filtrates, which contained a mixture of the different biodynes in addition to other substances, she placed large cakes of yeast in distilled water and exposed the cells to low levels of ultraviolet radiation. Afterwards, she removed the yeast cells from the liquid suspension by filtration. In order to assess the effect of the radiation on the yeast cells, she also prepared a control sample – a filtrate made from the same yeast under identical conditions as before, but not exposed to any damaging agents. She then photographed the UV spectra of both the irradiated and non-irradiated samples, plotted the graphs, and evaluated the differences between the two spectra. Her findings, which were published in 1938 in the prestigious journal *Nature*, showed that the crude filtrate from injured yeast cells had an absorption maximum at 2600 angstroms and a minimum at 2360 angstroms, similar to the spectra of nucleic acids and their derivatives.[53] She also performed a number of chemical tests of the filtrates by mixing them with various reagents. The conclusion of these tests was that the injured yeast filtrate, which possessed the biodynes, also contained DNA components (phosphorus, pentoses, guanine, and adenine)[54] – thus beginning Miriam's interest in the DNA bases.

Miriam's first published study revealed that the crude filtrate obtained from yeast cells damaged with UV radiation contained pieces of nucleic acids. These findings were later confirmed by Sister Mary Jane Hart, OP, who observed similar results while working with mechanically damaged yeast cells. Taken together, these results suggested to the researchers at the time that the biodynes were made up of nucleic acids. To test this idea, scientists at the Institutum isolated nucleic acids from different sources and tested them for growth-promoting activity. When tested, however, the isolated nucleic acids failed to induce cell division – leading researchers to suspect nucleic acid derivatives rather than the nucleic acids themselves.[55]

In addition to the biodyne study, Miriam recalled isolating nucleic acids for another reason. "We used to have a lot of arguments about what a *Euglena* was, because a *Euglena* was a one-celled organism which had chlorophyll, but it [also] had flagella. Motion is suppose to be a characteristic of an animal, and chlorophyll of a plant. And I had this naïve idea, that once I determined what its nucleic acid was, I would know. We saw yeast nucleic acid, as a prototype of what was in plants, and thymus nucleic acid, as the prototype of what was in animals. …I would take five pound cakes of baker's yeast, take the inner part out, and then extract nucleic acid out of that. Then I would go to the slaughterhouse, get thymus glands, and extract thymus glands to get nucleic acid. We had no idea [that] you would find in an animal both deoxyribonucleic acid [DNA] and ribonucleic acid [RNA]. We thought that they were separate, according to these two great classes of living organisms [animals and plants]."[56]

In the following years, Miriam continued to analyze the yeast filtrate, as well as different nucleic acid derivatives, by UV spectroscopy. Her UV spectrum of the crude yeast filtrate provided a fingerprint of the compounds present in the filtrate. In order to identify what the biodynes were, Miriam needed to find a compound with the same UV spectrum as that of the filtrate, similar to the way that one can identify a criminal by matching his fingerprints to those left at a crime scene. However, since UV spectroscopy was a relatively new method in those years, very little spectral information was available regarding the nucleic acid derivatives. Miriam spent the next two years obtaining the UV spectra of different nucleic acid derivatives and other cyclic hydrocarbons.[57] While the spectra of these compounds were similar to that of the crude yeast filtrate, none were a perfect match.

In the meantime, Cook realized a need to purify the biodynes from the crude yeast filtrate. He and his colleagues separated the crude yeast filtrate into different chemical fractions, which were found to stimulate the respiration of yeast and animal cells. Miriam and Sister Mary Jane further tested these fractions and discovered that some also stimulated the growth of yeast cells. Miriam then analyzed the samples by UV spectroscopy and observed that four of the five fractions, even those without growth-promoting activity, contained substances that absorbed light at 2600 angstroms.[58] These findings suggested that the ultraviolet absorption at 2600 angstroms was unrelated to the biodynes. The presence of contaminants in the samples would continue to complicate the characterization of the biodynes. Not until 1990 would researchers succeed in purifying the active yeast components down to a protein-containing fraction.[59]

In 1939, Miriam and Sister Mary Jane Hart, OP, received their Masters of Science degrees and graduated from the Institutum Divi Thomae. Best friends for the better part of three years, the paths of these two women then took them into different directions. Sister Mary

Jane was sent to West Palm Beach, where she taught at Rosarian Academy and continued scientific research at the Florida branch of the Institutum Divi Thomae. In 1940, the Adrian Dominican Sisters opened Barry College, and Sister Mary Jane joined as part of the first faculty. There, she then spent the next nineteen years teaching chemistry.[60] Miriam, on the other hand, returned to Adrian, Michigan, where she joined the faculty at St. Joseph College, which changed its name to Siena Heights College that year.

More than sixty years later, Miriam's work at the Institutum was remembered in Adrian in the form of a local urban legend.[61] In order to help finance the cancer research at the Institutum, Dr. Sperti licensed the sale of a biodyne-containing ointment as a wound healing agent for minor burns and abrasions. Contrary to the urban myth, Miriam did not hold the patent to this ointment. However, her role in the biodyne research will forever be remembered by the trade name of the ointment – Preparation H.

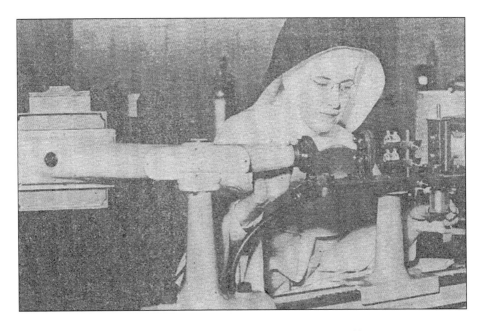

Sister Miriam Michael Stimson, OP, using a Hilger quartz spectrograph in conjunction with a Spekker photometer.

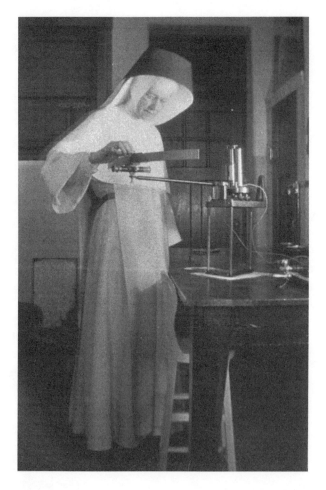

Sister Miriam Michael Stimson, OP, with a galvanometer.

Chapter 7

Miriam's Children

*Among my fellow classmates, she (Sister Miriam Michael) had a nick-
name of M^2, derived of course from the initials of her name, but also
signifying her intellectual brilliance.*
— Julia Franko Opalek

F ollowing the instructions of Mother Gerald, Miriam returned to
Adrian, Michigan after receiving her Master of Science degree
and joined the faculty of Siena Heights College, formerly
known as St. Joseph College. At that time, the faculty was comprised
entirely of Dominican Sisters, who affectionately referred to the all-
female student body as their children. As a member of the faculty,
Miriam taught approximately three science courses each semester,
which included lectures and laboratory sessions, as well as advised students on their academic schedules. In addition to her teaching duties,
Miriam was also a "wing-nun," a sister who resided in the dormitory
and was responsible for the welfare of the young women who roomed
on her floor. Furthermore, Miriam also fulfilled her obligations to the
Institutum Divi Thomae by establishing a branch research unit at Siena
Heights College.

When Miriam joined the faculty of Siena Heights College, she
found herself reunited with Sister Agnita [Grace] Reuter, OP, her former chemistry teacher and now colleague. Similar to Miriam, Sister
Agnita had been born in Chicago to parents of Irish Catholic descent.
Although unlike Miriam, her parents were not fervent Catholics. Instead, Sister Agnita's father possessed a hatred of anything Roman
Catholic.[1] Due to her father's disposition, Sister Agnita's mother raised

her in the Methodist belief. However, when Sister Agnita was 12 years old, she and her mother were reunited with the Church, two years before her mother passed away. Later, at 18 years of age and against the wishes of her father, she joined the Adrian Dominican Sisters. While Miriam always enjoyed the support of her family in her religious calling, Sister Agnita did not. According to Sister Agnita, "I suffered for my faith. No one from my family came to my reception or profession or visited me."[2] Although scorned by her father, she was proud of her heritage and told many of her relation with the family that owned the Reuter news agency.[3]

As a new teacher, Miriam shared many qualities with Sister Agnita. Miriam had never taught before arriving at Siena Heights College, and she quickly developed a teaching style that mirrored Sister Agnita's. Sister Agnita was a very strict and forthright instructor. Most people who knew her would agree that she was not a subtle woman. In standing up for what she believed was right, she was self-assertive and even brash at times. However, under her hard shell, people found Sister Agnita to be contrite and even docile.[4] Likewise, Miriam was remembered as an imposing teacher who demanded nothing less than the best from each of her students. Many of the science students found a caring, lifelong friend beneath Miriam's formidable exterior.

Some of Miriam's former students first met her when they were high school seniors visiting the campus. One such meeting was recalled by Diana Albera Luciani. "I remember my first encounter with her [Sister Miriam Michael] very well. She was a tall, handsome woman – a very imposing figure! When Sister Kevin introduced me as a prospective pre-med student, there was a slight, but very visible change in her demeanor. She began to question me at length about why I wanted to study science in general and medicine in particular. She made it clear right from the beginning that medical school was an extremely difficult place for a woman. I believe Donita Sullivan, a recent Siena alumna, was still attending medical school, and Sister used some of Donita's horror stories to illustrate her point rather forcibly. Sister pulled no punches when she told us that she gave her science majors a hard time and made life impossible for her pre-med students... Life with her would be as miserable as possible because the environment for women in medical school was less than ideal. You had to be really tough to survive... There was no doubting her sincerity, but, strangely enough, she didn't frighten me or even make me nervous at that time."

For other students, their first encounter with Miriam came when they registered for classes. Dorothy Nyhoff recalled the following:

"She asked me what I wanted to major in and I replied, 'chemistry.' Then she asked me if I had taken chemistry in high school and I said, 'no.' She asked if I had taken physics in high school and I said, 'no.' She asked if I had taken trigonometry in high school and I said, 'no.' She said, 'Let's go back to the beginning again. What do you want to major in?' After a few more minutes of conversation, she registered me for the beginning chemistry class. For the next four years, she did her best to remember that I was starting from square one and always began with the basics, even if other members in the class already knew some of the material. I know I would not have survived in a harsher environment." At another advising session, Ursula Ording, OP, recalled, "Despite our intended major, she [Miriam] convinced most of us that it should be chemistry. My math and English teachers weren't happy about that." "She always *told* us what subjects we were taking each semester," remarked Christine Kazen Cauchi, another former student.

Miriam also made a memorable impression on the first day of classes. One of these occasions was recalled by Donita Sullivan. "I remember my first class in chemistry freshman year when she loudly proclaimed to the class that if there was anyone there who had signed up for an easy class in order to meet their science requirement that they should leave now!! The next day the class numbers had shrunk by a recognizable number." A similar event was experienced by Diana Albera Luciani. "Sister took a perverse delight in telling her freshman chemistry class that most of them would change their majors by the end of the first semester!"

Many of Miriam's former students affectionately recalled Miriam's intimidating intellectual presence. One such student was Joanne Hettrick. "I remember her, initially, as a strong woman of great intellect with a somewhat forbidding demeanor. Many of the young student nuns in the chemistry department were somewhat in fear and awe of her. She was a no-nonsense woman who took her academic and religious life seriously. She guided her chemistry majors wisely and with a firm hand. Often, she had us taking very heavy class loads in which we were expected to achieve a 'B' or better to stay in the chemistry department. The rumor was that Miriam was one who had her own way of determining those who would succeed in the department: a 'survival of the fittest' (or the most stubborn)." Similar feelings were echoed by Sister Marie Siena, OP. "I was completely awed by her. At first I was frightened because I didn't think I could come up to her standards, but as I began to know her, I realized that her students were of uppermost importance to her." "I loved her and ... feared her," summarized Sister Mary Alan Stuart, OP.

In the classroom, Miriam was as demanding as she had promised,[5] and she soon earned a reputation as a tough and exacting teacher. "Sister Miriam was a challenging and stimulating teacher," wrote Sister Helen Duggan, OP. "She demanded serious and consistent effort from her students but always extended herself in help and support as needed." Sister Therese Mary Foote, OP, had a similar remembrance. "Sister Miriam Michael was a strong, strict teacher, very knowledgeable and always well-prepared. She demanded the best, and at times, even our fear of her sometimes. [Her] stern approach was a strong incentive to study harder. When Sister Miriam gave a compliment, it was indeed an honor!" Sister Sharon Weber, OP, recalled, "Miriam once said – as she was challenging us as students – that a good teacher did not aspire to help her students be as good as she was, but to help them be better than she was. She took quite seriously her responsibility to influence the future through providing excellent preparation for the next generation. And she never stopped doing that – challenging the next generation." "She challenged and drew the best out of students," summarized Sister Bernard Lynch, OP.

While Miriam's strict teaching style may seem stringent by today's standards, her tough-love approach was what some students needed. Donita Sullivan recalled, "From the outset of my college career, she [Miriam] had been very negative about the probability of my success (being in pre-med) getting into medical school since so few women were admitted in those days. I should therefore, plan a major that would allow going to grad school as the appropriate alternative. So I did a double major in chemistry and zoology but did not give up on applying to med school (I was going to show her!!). Much to everyone's surprise, I was accepted by 3 different medical schools. All along, she always knew which of my buttons to push and how I would respond. Her methods of holding her students to high standards, not giving us anything more than what we earned, prepared me well for medical school. The other woman in my medical school class (there were only the 2 of us in a class of 125) came from a college program where grades were determined on a curve. She had a very difficult adjustment in med school as compared to what I had experienced at Siena. In medical school, Miriam kept in touch and was always very supportive. Sometimes, even there, she would prod me a bit by questioning why I had not gotten a better grade. Once again I had 'to show her' so I graduated second in my class from St. Louis University Medical School!!!'"

As the students became better acquainted with Miriam, they discovered a gentler side of her. "Miriam kept a rather ancient two burner gas 'stove' in her lab, at the end of one of the counters," described Joanne Hettrick. "Miriam was a tea lover and would brew her tea on those burners. One day, she offered me a cup of tea. It was, in her words, 'high-fermentation tea.' That was the most interesting cup of tea I had ever had. Interesting in flavor and interesting in the aspects of Miriam's personality I discovered that day. I don't even remember what we discussed, but I remember seeing the Miriam [that] I had always felt was there under the tough exterior. No, she wasn't 'soft', but she was a real human being who cared deeply about her students and their success and happiness." Similar feelings were shared by Diana Albera Luciani. "She was a tall, imposing presence, but we soon learned that she respected hard work and determination, and that she had a softer side surfacing just often enough to keep this fearsome image she projected from frightening away all of her perspective students!"

Miriam's gentler side was also revealed through her support of the liberal arts. According to a former student, "Sister Miriam Michael believed very strongly in turning out well-rounded science majors. She often commented that most scientists could only carry on an intelligent conversation about science."[6] Consequently, Miriam insisted that her students take classes in subjects outside of science, such as in art appreciation, literature, and sociology. A student recalled, "She made it very clear that she had little use for one-sided education – it made for very dull conversation."[7] Miriam also loved classical music and encouraged its appreciation by often playing recordings of classical music during the laboratory sessions. A student summarized, "She was always eager to share her knowledge, to make you aware of many facets of a situation, to be curious about everything, to ask questions, and to be interested in life."[8]

Miriam taught a number of subjects at Siena Heights College, but mostly lectured on chemistry. When she began teaching, the science lecture rooms and laboratories were located on the fourth floor of Sacred Heart Hall. Miriam possessed a distinctive, nasally voice, and when she lectured, "You could hear her voice all over the fourth floor," recalled Sister Mary Beaubien, OP. "Not that her voice was loud – it just carried." "These lectures often were in late afternoon when it was hot and humid," noted Sister Mary Alan Stuart, OP. "No matter how interesting Sister Miriam's lectures were, invariably someone nodded off. As her watchful eye caught the offender, her voice would reach a slightly higher pitch and she would call out the name of the person. Sister then invited the student to come forward, grab a hard candy or two from the jar on her desk, and stand a while at the back for a better view of the chalkboard. We all would straighten up and be more attentive."

Miriam also taught the upper-level science courses, which sometimes only had two students. Since there were so few chemistry majors, Miriam often held class in her office. "I remember sitting across from her at her desk," recalled Dorothy Nyhoff. "She would pull out this spiral notebook with color-coordinated notes and explain away. Try not doing your homework with a class of two! I don't remember ever being frustrated or put down. She was a wonderful teacher and I didn't appreciate her enough until several years later when I took courses at a local college and found out how good she was." Jeanne Sheck had a similar experience. "Sister Miriam Michael … was an excellent teacher. She was always well organized and well prepared. In the upper level courses of my senior year, the classes consisted of two students. She prepared as well for two students as she did for a full classroom. An example of her dedication as a teacher: she destroyed her organic chemistry lecture notes at the conclusion of each course in order to keep the following year's lectures fresh."

In addition to chemistry, Miriam also taught biochemistry. "Because the class was small, we often had lectures in her lab," remembered Diana Albera Luciani. "She would even prepare her famous tea occasionally! Most of the other students (if not all of the other students) were studying for a degree in Medical Technology. This meant that some of the practical aspects of biochemistry would be applied in class! Two memorable lessons involved drawing blood and stomach pumping. Sister was the guinea pig for the blood, and all of the students had to draw blood from her once. Her theory was that if we could do it with her, and then to each other, we could do it with anyone. You can imagine the case of nerves we all had on the appointed day. We all managed to get through it, some easier than others. The stomach pumping was even worse, however. It meant that we had to swallow a tube, place the other end with the pouch into our lab coat pocket, and attend our morning classes in this state. Because it was a yearly ritual, the teachers were accustomed to seeing these poor creatures struggling to behave normally, but the other students were another matter. I can recall very vividly all the stares, snickers, and comments. Most of them felt sorry for us, the rest thought we were just plain weird. We all survived, and believe me, we knew exactly how it felt to have blood drawn and stomachs pumped. I guess if it served to make us more sympathetic to the poor patient, then there was a method to Sister's madness!!"

Miriam was also remembered as an innovative instructor, who incorporated modern instrumental methods into the courses she taught. While at Siena Heights College, Miriam recognized that the classical teachings of chemistry failed to prepare the students for the realities of modern laboratory practice. In response, she emphasized the use of scientific instruments as part of the regular chemistry curriculum beginning with the freshman year. As Miriam gradually acquired new instruments for her scientific research, she selflessly permitted students to operate them. Diana Albera Luciani gratefully wrote, "We all came to appreciate more and more the kind of training we were receiving at Siena, not to mention the fantastic equipment Sister Miriam Michael had and actually allowed her students to use."

For some students, Miriam's best teaching was done outside of the classroom, by her example. "Sister Miriam was a brilliant teacher and motivator," remembered Sister Mary Alan Stuart, OP. "Likewise, I knew that I could learn a lot from her by being around her. So, one summer I volunteered to do odd jobs in her research lab and office area. I though she might assign me to some special project. Not so … she had me washing all the glassware that piled up. One day I broke the top of an expensive long-necked flask. I humbly admitted my crime, asking where to discard the item. Sister Miriam, not one to waste anything, directed me in the task of trimming down the rough edges and smoothing it off with a piece of coarse sanding cloth. I used this trick many times in my years of teaching." Diana Albera Luciani shared a similar sentiment. "My summer at Siena working with Miriam Michael was surely one of the most exciting periods in my life."

Over the years, Miriam's demanding, yet caring, personality made her one of the most revered teachers at Siena Heights College. "People found her love of science, research, and learning in general to be contagious," observed Judith Redwine. She was "demanding enough, but always helpful," noted Sister Anna Rita Sullivan, OP. Sister Marie Siena, OP, wrote, "I regard Sister Miriam Michael as one of the best teachers I ever had. She not only taught me content, but motivated me to do my best." "We all had a tremendous respect and affection for her," recalled Dorothy Nyhoff, "she was a teacher that cared about the entire person." Jeanne Sheck remembered, "In my student years, she was readily available to counsel a worried student. It was from Sister Miriam Michael that I learned a valuable lesson – failure is acceptable if one has utilized one's gift/talents to the maximum; that failure is not reflective of one's worth as a person." Likewise, Karen Erickson recalled Miriam saying, "You have not matured until you have failed at something." Ursula Ording summarized, "She was a teacher you never forget."

In addition to her teaching duties, Miriam was also a wing-nun, a sister who lived among the students in the dormitory. Like a mother, Miriam looked after the young women who roomed on her floor. She offered them assistance and advice, made sure that they kept their rooms in respectable order, and reminded them of the various rules and regulations.

One former student had the following memory of Miriam and life in the student dormitory. "Freshman year, my room was located next to one of the parlors. My window faced Walsh Hall, and I could hear the music students practicing during the warmer months. It is something I remember with a great deal of pleasure. There was a large statue of the Sacred Heart at the end of the hall. I can still recall walking back from evening classes or the lab and seeing the red vigil light glowing in front of the statue… The chapel was just inside of the main entrance [of Archangelus Residence Hall] until my junior year when Lumen [Ecclesiae Chapel] was completed. The old chapel was turned into a lounge. I liked having the chapel just a few steps away from my room. I used to leave my door open and listen to the nuns praying and singing… One of the most interesting things for me was actually living in the same area as the teachers. I had two nuns living across the hall from me freshman year. I had been around nuns all my life, but never in such close quarters. The experience made it very clear that they were human beings just like the rest of us. They got headaches and cramps, they liked to visit and share goodies, and they were good listeners if you had personal problems. They got mad at each other, and lost their cool occasionally with the students. Back in the 'old days' there was a tendency not to think of the religious as human beings. These women were not afraid to be themselves, and I respected them all the more for it… Living in such close proximity to the faculty brings to mind the fact that the upperclassmen loved to trick the freshmen into doing some pretty odd things. One prank I have never forgotten involved a close friend of mine and Sister Miriam Michael! Sister's room at that time was located next to the incinerator door. My friend had little regard for rules and stayed up past 'lights-out' all the time. She would then sneak down to the room of a junior friend to play cards or listen to records. One particular late night, she decided to empty out her wastebasket. She wasn't in the habit of doing much cleaning – her roommate took care of chores, so she didn't know where the incinerator was. She stopped to ask her friend who, of course, pointed out Sister's door. Her friend told her just to open the door and toss in the garbage! She flung open the door, none too quietly, and was about to empty her trash when she noticed a white habit hanging over a chair."[9] "Apologizing profusely and stammering an explanation, she slammed the door and ran to her room! The volcanic eruption the prankster expected didn't occur and they never figured out if Sister was asleep or just a good sport!"[10]

Another former resident of the student dormitory had this recollection of Miriam: "The story that comes to mind concerns bedbugs. Sister [Miriam] was monitor of the Archangelus dormitory wing where I roomed. My roommate and I noticed a number of small marks on our arms and legs that looked like insect bites. We voiced our concerns about bedbugs to Sister. A flurry of activity erupted the following Saturday. Under Sister's direction, every bed in the wing was stripped, moved to the fire-escape landing, sprayed and returned to clean bed linens, as well as clean rooms. We found no evidence of bug infestation. My roommate and I certainly found wing-mates who were very annoyed with us. Fortunately, the day was a warm spring day."[11]

In addition to her teaching and wing-nun duties, Miriam also established a branch research unit of the Institutum Divi Thomae at Siena Heights College. Miriam explained, "Well the idea was that after we had been trained, when we went back to our home colleges, the college was to set up a unit with the Institutum Divi Thomae. Dr. Sperti had the idea that every place should be doing something with mice, as a visible sign that we were working with cancer."[12] Although the science classrooms and laboratories were located on the fourth floor of Sacred Heart Hall, the only available space for her research laboratory was on the first floor. There, Miriam was allowed to convert a spare bathroom into research space, and the toilets and wash basins were soon removed. The modest-sized room was then subdivided. One part was set-up as a darkroom for developing photographic plates. Another section was designated as the animal room for housing the lab mice. Yet another area contained a large still for making distilled water. With little room left for storing chemicals, the bulk of the chemical reagents used in her research and teaching were stored on shelves mounted along the stairwell that led down to the sub-basement. Since the building lacked an elevator at that time, Miriam often had to carry chemicals from the sub-basement or carboys of water from the first floor still up the stairs to the fourth floor teaching laboratories. "A carboy itself is very heavy. With five gallons of water in it, that is pretty heavy for women," noted Miriam. "We got our exercise that way."[13]

To continue her scientific research, Miriam also had to secure financial resources. When Miriam set-up her research laboratory in 1940, the country was still recovering from the 1929 stock market collapse, which had left millions without work, savings, or a home. With the nation still in the midst of intractable poverty, the Adrian Dominican Sisters, too, found their resources stretched to the limits. As a result, the Congregation and the College were unable to offer Miriam any monetary support for her research. "The lack of money was really a critical issue," remarked Miriam.[14] "From the very beginning, there was a problem with what we would call nowadays, 'budget.' We did not have a research budget. I would make attempts to get funds from one thing or another, like my parents. My father and mother would donate, but that was just a bit."[15] Consequently, Miriam had to go out and raise the money to support her work. In an era before large federal government agencies, like the National Cancer Institute, National Science Foundation, and the National Institutes of Health, Miriam focused her fund raising activities on the county branch of the American Cancer Society, local businesses, and private donors.

Miriam's responsibilities as an Adrian Dominican sister, teacher, wing-nun, and fund-raiser, left her with little time to conduct her research. She explained, "When we set up the [branch unit of the] Institutum. I was supposed, theoretically, to have half time [off] for research: that was the idea. That time didn't exist. People at the college never knew anything about that. So you had all your obediences [obligations to the Adrian Dominican Sisters], all these classes, lots of things."[16] Over the next decade, the College did little to alleviate Miriam's workload. "There is so much to do and so little time to do it," she told a reporter in 1951.[17] "You had to steal the time [for research]... It wasn't built into the day."[18] The hectic schedule consequently made it difficult for Miriam to even think about her research. Miriam explained, "You were always turning psychological somersaults. You would teach for an hour or an hour and a half, then research for an hour or so, then teach, [and] then research."[19] As a result, Miriam performed much of her research on the weekends or over the summer, when classes were not in session.

Although Miriam struggled to find sufficient time and funding for her work, scientific research became an integral part of her life. Her research was the common thread that unified the Dominican sister, teacher, and scientist in Miriam, and helped her to become more complete in each of these facets of herself. To Miriam, research was her avenue of worship, through which she could express herself as an Adrian Dominican sister in the search for truth. Research also enabled Miriam to extend the deposit of knowledge and to make a contribution not only in the classroom as a teacher, but to humanity as a scholar. Furthermore, since the religious sisters lived rather regimented lives, research provided Miriam with an important sense of freedom, with which she could explore the endless realm of the intellectual.

Once the new research unit was formed, Miriam continued her UV studies of the DNA bases. As a graduate student, Miriam studied the ultraviolet absorption of the purines (adenine and guanine) and pyrimidines (cytosine and thymine) because of their association with the biodynes and cancer. However, during the accumulation of this information for biochemical purposes, Miriam developed an academic interest in the spectral qualities *per se* of the purines and pyrimidines.[20] At the Institutum, Miriam used a Hilger quartz spectrograph in conjunction with a Spekker photometer, which entailed photographing the UV spectrum and plotting it by hand. Using this instrument, she published a series of papers on the UV absorption spectra of a number of purines and pyrimidines, starting with adenine[21] in 1940 and uracil[22] in 1943. Later, using one of the first commercially available spectrophotometers, a Beckman model DU, she published the UV absorption spectra of thymine,[23] cytosine,[24] and guanine[25] in 1945.

During the initial years of her research career, Miriam was one of the first to apply a method called paper chromatography to the study of the DNA bases. Paper chromatography is a separation method used to analyze the components of a complex mixture. Initially used in the 1800s, the dye industry employed a crude form of paper chromatography to assess the purity of their dyes. By spotting vat solutions on paper and then observing the number of concentric rings formed, they could estimate the number of components in their dyes. Although paper chromatography has had a long history, its usefulness for studying organic compounds was not fully realized until 1944, when amino acids were successfully separated by this method.[26]

Paper chromatography is a relatively simple procedure that utilizes filter paper and an appropriate solvent. Usually, small spots of dissolved sample are applied near the end of a strip of dry filter paper. The end of the paper is then dipped into a solvent, which is slowly absorbed by the paper. As the solvent gradually migrates up the length of the paper, the components in the sample are carried along with it. These substances, however, often have different affinities for the solvent and paper, causing them to travel up the paper at different rates. The result, called a paper chromatogram, is the physical separation of the components along the length of the paper, which can then be detected, for example, by their fluorescence under ultraviolet light.

Miriam used paper chromatography to assess the purity of her DNA base preparations. If her samples contained impurities, they could be separated from the DNA bases by paper chromatography and subsequently detected as extraneous spots on the filter paper. Miriam recalled, "Some of the purines and pyrimidines I would have to synthesize. You could buy a few, but not very many. You had to make your own. One of the attempts that I was making was because I wanted to run spectra on them. The hope was that if we could get the spectra of the identifying parts, we would be able to identify [them] in the cancer cell. Well, in order to measure spectra, either ultraviolet or infrared, they would have to be very pure."[27] "I was using paper strips for chromatography, because I would have to purify the chemicals that I was using, particularly with the thymus nucleic acid. It was very crude. So I was using these papers and dipping them into solutions and seeing how far they raise, would come-up, and then I would use an ultraviolet light on them to see where there were things that showed up... I should have called that paper chromatography," lamented Miriam, who missed an opportunity to publish on this aspect of her work.[28]

Although Miriam did not publish a manuscript on the use of paper chromatography with the DNA bases, she did co-author a paper in 1941 on the fluorescence of these compounds.[29] In her article published in the *Journal of the American Chemical Society*, she described the color and intensity of various purines and pyrimidines, including uracil, cytosine, guanine, and adenine, that were exposed to a Sperti UV lamp. Miriam also presented these findings at a meeting of the American Chemical Society, which was briefly described by a local newspaper reporter. "...they [the attendees] were told... by two Catholic Sisters that sick organs of the body can be detected by a yellow color that shows under 'black' or ultraviolet light. Sisters Miriam Michael Stimson and Mary Agnita Reuter of Siena Heights College, Adrian, Mich., said the yellow color comes from substances known as purines and pyrimidines."[30] Miriam's pioneering work with UV spectroscopy, paper chromatography, and UV-induced fluorescence laid the foundations for

Erwin Chargaff, who approximately six years later successfully used these methods to quantify the amounts of each of the DNA bases in a historic experiment.

After Avery, Macleod, and McCarty published their work demonstrating that the transforming substance from *Streptococcus pneumoniae* was DNA, Erwin Chargaff, an Austrian-borne biochemist working at Columbia University, realized the importance of DNA and decided to devote his research to this molecule.[31] Chargaff wanted to address the tetranucleotide hypothesis, the belief that the four DNA bases were present in equal proportions. However, the techniques that were employed at that time to isolate and hydrolyze DNA often resulted in the chemical modification or the destruction of the DNA bases. This was the reason why Levene and Steudel had obtained contradictory results in 1905 and 1906.[32] Realizing this problem, Chargaff's research group spent years developing gentler techniques of isolating DNA and cleaving the chemical bonds that joined the nucleotides. Afterwards, they needed a method of separating the DNA bases, and they then turned their attention to paper chromatography.

With knowledge of the successful application of paper chromatography to separate the amino acids of proteins, Erwin Chargaff and Ernst Vischer decided to adapt this method for use with the DNA bases.[33] After hydrolyzing their DNA samples, they used paper chromatography to separate the DNA bases – adenine was separated from guanine, thymine from cytosine, and so on. At first, they tried to locate the purines and pyrimidines on the paper chromatograms using an ultraviolet lamp as Miriam had done. Whether their decision to use a UV lamp was based on Miriam's 1941 paper is uncertain. Their initial attempt to observe the fluorescence of the DNA bases, however, failed due to problems with their Hanovia lamp.[34] Instead, they had to resort to a more tedious procedure of simultaneously processing two paper chromatograms – using one for the chemical detection of the DNA bases and the other for analysis. Later, though, they abandoned such duplications, and used a short-wave UV lamp to detect the separated purines and pyrimidines more easily.[35]

Once the DNA bases were effectively separated, Chargaff and Vischer needed a method of quantifying the amounts of each of the DNA bases. In 1948, their efforts then turned to UV spectroscopy. After they eluted each DNA base from their locations on the filter paper, they then analyzed the samples by UV spectroscopy. Using molar extinction coefficients calculated from the UV absorption spectra, as Miriam had done as early as 1940, Chargaff and Vischer were able to quantify the amounts of each of the DNA bases in their samples.[36] Although Vischer and Chargaff did not cite Miriam's work directly in their 1948 paper, they did refer to the methods used by H. M. Kalckar,[37] who cited one of Miriam's spectral studies in his paper. By using such methods, Vischer and Chargaff discovered from the DNA of different organisms, like animals, fungi, and bacteria, that the amount of purines (adenine and guanine) was equivalent to the amount of pyrimidines (cytosine and thymine). Furthermore, the proportion of adenine was equal to thymine, while the proportion of guanine was equivalent to cytosine.[38] This relationship between the DNA bases, which was published in 1949, would later be known as Chargaff's rules or ratios.

Chargaff's work was significant because it demonstrated that the tetranucleotide hypothesis, which had become dogma for approximately forty years, was false. DNA was not the simple repetitive molecule as previously thought. Instead, Chargaff's work showed that the proportions of the four DNA bases varied from one species to another, implying that DNA was as diverse as life itself. His data also revealed that different combinations of the four DNA bases could encode an enormous quantity of information just as a protein could with twenty amino acids. For example, Chargaff calculated that the number of DNA base combinations equivalent to that of an ox were 10^{56} for a chain of only 100 nucleotides and 10^{1500} for a chain of 2,500 nucleotides.[39] In the minds of scientists, these results helped to convince them that DNA, and not protein, was the genetic material.

Chargaff's ratios also had important implications for the structure of DNA. Although the amounts of the four DNA bases varied from one organism to another, the proportion of purines was equivalent to that of the pyrimidines for all species. Furthermore, scientists would have to account for Chargaff's ratios (adenine = thymine, guanine = cytosine) when building models of DNA. Although regarded as the most important development in the chemistry of the nucleic acids of that time, the meaning and importance of this relationship was not fully realized.[40]

Sister Miriam Michael Stimson, OP, adjusting the pH
of one of her solutions.

Chapter 8

Tautomerism

It sounds paradoxical to say the attainment of scientific truth has been effected, to a great extent, by the help of scientific errors.
– Thomas Huxley

In 1942, the largest nations of the world were once again at war. Axis powers were sweeping through Europe, Africa, and Asia, and scientists from all nations were called upon to serve their countries by conducting war-related research. While many scientists were engaged in wartime weapons research, such as the Manhattan (atomic bomb) project, most female scientists, including Miriam, did not participate in such activities. Instead, Miriam made her patriotic contribution by studying quinine, the anti-malaria drug. Quinine is a natural product found in the bark of the cinchona tree, which grows in tropical rainforests. When Japanese forces invaded Java in 1942, the Allies' supply of quinine was cut off. Since hundreds of thousands of Allied troops needed quinine to battle Japanese troops in malaria-prone Southeast Asia, a new source of the anti-malaria drug was badly needed. To address this problem, Miriam recorded the UV spectra of quinine and a number of other cinchona alkaloids.[1] Her data then provided scientists with a method of estimating the amount of quinine that could be extracted from various barks, thereby giving researchers a tool to search for new sources of quinine. Miriam's interest in quinine was short-lived, however, and ended with the cessation of World War II. Throughout the war years, Miriam's primary scientific focus remained on the DNA bases.

In the spring of 1944, a Beckman spectrophotometer, model DU, became available and Miriam was among the first to work with this new instrument. Miriam recalled, "Mother Gerald gave the money for us to get a Beckman UV Spectrophotometer. Before that [time] there was no photoelectric way of doing these things [UV spectroscopy]. Before that, the method of obtaining spectra was to use a split beam [spectrograph], and to photograph the spectra after the beam went through the sample and through the solvent, and comparing them visually. That was pretty hard on the eyes. You had to develop the photographic plate and make an enlargement of that, generally onto paper, and then try to read it. If you were sophisticated, you tried to get a photometer, otherwise you used your eyes... The Beckman [spectrophotometer] was an advance, because you [the researcher] had numbers read off, so you could plot the wavelength, the numbers, versus the absorbency. That was the first kind of thing that could be done that way, with, shall we say, quantitative aspects."[2] Using the most advanced UV instrument of her time, Miriam continued her studies of the DNA bases.

Much of Miriam's research at that time was focused on the effect of pH on the structures of the purines and pyrimidines, which were inferred indirectly by their UV absorption spectra. During this work, she recorded the UV absorption spectra of numerous chemical compounds over a range of pH values. In one study, for example, she prepared three different solutions of thymine of varying pH values (pH 3, 7, and 11). She then recorded the UV absorption spectra of each of the three solutions and plotted them on an overlay graph. As shown below, she found that raising the pH of thymine caused a shift in the absorption maximum, the UV wavelength with the greatest absorbance, toward a higher wavelength. In addition to thymine,[3] she reported similar results with adenine,[4] uracil,[5] cytosine,[6] and guanine.[7]

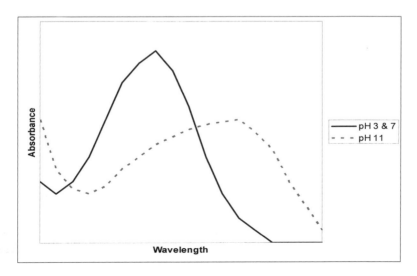

Miriam interpreted her experimental results in light of prior studies. Previously, Francis Heyroth and John Loofbourow observed that
raising the pH of uracil caused the UV absorption maximas to shift
toward the longer wavelengths. They suggested that pH induced a
structural change in uracil, which they believed was attributed to
tautomerism. Tautomers are structural isomers (different arrangements of the same atoms) that vary in the position of a hydrogen
atom. The different tautomeric forms are believed to rapidly interconvert so that they exist together in equilibrium, similar to
resonance. Heyroth and Loofbourow believed that uracil existed in
both the keto and enol tautomeric forms. The differences they observed at low and high pH, they thought, were due to uracil "going
from the lactam [keto] to the lactim [enol] form."[8]

Consistent with Loofbourow's conclusions regarding uracil, a
pyrimidine found in RNA, Miriam suggested that tautomerism also explained the effects of pH on the UV absorption spectra of the DNA
pyrimidines, thymine and cytosine. She wrote, "The marked change in
absorption with change in pH is consistent with the [tautomerism] theory …"[9] In general, Miriam believed that the pyrimidines "existed in
the keto form in acid solution and that enolization [the switch from the
keto to the enol form] does not take place until well above the neutral
point."[10] In other words, she thought that thymine, for instance, would
exist in the keto configuration at pH 3 and 7, while the enol form would
prevail at pH 11.

keto   enol

Similar to the pyrimidines, Miriam also believed that tautomerism
explained the pH effect on the UV spectrum of adenine, a purine. She
wrote, "Examination of the ultra-violet spectrum of adenine over a
wide range of pH showed … that the spectrum of this compound in
aqueous solution is appreciably influenced by pH… We believe the
changes in absorbance observed to be attributable to a tautomeric rear-

rangement ..."[11] In the case with purines, Miriam believed that the amino tautomer would exist at low pH, while the imino form would occur at high pH.

amino   imino

The 1940s were the most prolific years of Miriam's research career. During that period, she authored fifteen papers, eleven of which dealt with the effects of pH on the UV absorption spectra of the purines and pyrimidines. Furthermore, in 1948, she completed her doctoral thesis on the molecular symmetry of the pyrimidines and the effects of pH on their UV absorption spectra, and she was awarded a Ph.D. from the Institutum Divi Thomae.[12] In 1949, the gist of her thesis work was published in the *Journal of the American Chemical Society*.[13] At that time, Miriam's research career appeared to be going extremely well. She was averaging more than one new finding each year, she was working with one of the latest scientific instruments, and she had just earned her doctorate. But then, her pH studies came to a sudden halt, and her career took a sadder turn. At that moment, Miriam realized what every scientist secretly fears – she had misinterpreted her results and had published a mistake.

In 1951, J. R. Marshall and James Walker concluded that the shift in the UV absorbance spectra of the purines was an ionic phenomenon rather than due to tautomerism (enolization). They critically wrote the following of Miriam's work. "The value of some of the previous work carried out on aqueous solutions must really be regarded as slight since the hydrogen-ion concentrations at which ultra-violet light absorption has been measured have frequently been chosen without either knowledge of, or any reference to, the pKa values of the substances being studied, with the result that absorption measurements have often been made on mixtures of ions and neutral molecules, and changes in the shapes of extinction curves with changes in pH have been attributed to 'enolization,' whereas ionization has really been the explanation of the effects observed."[14] Ionization is a process by which atoms become positively or negatively charged. Normally, the DNA bases are uncharged, as shown below with thymine and adenine. However, at high pH, thymine can lose a positively-charged proton and become negatively-charged, while at low pH, adenine can gain a proton and become positively-charged. Hence, the shift in the UV absorption spectra, which Miriam had attributed to the different tautomers, was actually due to the presence of the charged forms of the DNA bases.

Thymine (uncharged)   Thymine (charged)

Adenine (uncharged)   Adenine (charged)

In addition to having her work publicly lambasted by Marshall and Walker, Miriam also had to endure the patriarchal attitudes possessed by some of the male chemists. Miriam was a woman in a male-dominated field, and her mistake was not unnoticed by her misogynistic male counterparts. Dorothy Nyhoff, a former student, shared the following recollection. "I remember her sharing a story once about being at a conference and a man coming up to tell her that he could prove what she had once written was wrong. She now knew it was, but graciously let the individual give his explanation. She warned us that it was hard to take back anything that had gone into print." Miriam never made excuses for her mistake, but rather gracefully accepted it, although regretfully.

It is not uncommon for scientists to misinterpret their data. For instance, Phoebus Levene mistakenly advanced the tetranucleotide hypothesis, which adversely affected the way scientists viewed the structure and function of DNA for approximately forty years. Rosalind Franklin stubbornly believed that the A form of DNA was not helical, although it was, and even Linus Pauling authored an erroneous paper that promoted a model of DNA that contained the wrong number of sugar-phosphate strands. While Miriam's misstep may seem minor in comparison to those of others, several factors played into compounding the error.

One important factor that magnified Miriam's mistake was that she published eleven papers on the effects of pH and propounded the idea of tautomerism for nearly ten years. Although others also explained their experimental results in terms of tautomerism,[15] Miriam was the most prolific author of this group of scientists. So in many respects, her success in publishing so many research articles perpetuated the belief in tautomerism.

The topic of Miriam's work also played a significant role in compounding her mistake. Miriam's pH studies were not just conducted with DNA, considered by many to be the most significant biological molecule, but with the most important part of DNA – the bases. Miriam's experiments were performed during the era when scientists were just beginning to grasp the importance of DNA as the genetic material. If she had focused her pH studies on the esoteric cinchona alkaloids, on which she published a paper in 1946,[16] instead of the purines and pyrimidines of DNA, then her erroneous interpretation would most likely have been forgotten by history.

Furthermore, the instruments employed at that time were not sufficient to unequivocally determine the structure of the DNA bases. More powerful methods of chemical identification, such as mass spectrometry and nuclear magnetic resonance, were not available at that time. Even x-ray diffraction failed to yield definitive answers. In 1956, for example, Linus Pauling summarized, "Not many reliable x-ray determinations of the structure of crystals of these substances have been made, and at the present time only tentative conclusions can be reached as to the probable interatomic distances and bond angles in the residues of adenine, thymine, guanine, and cytosine in the polynucleotide chains of DNA."[17] Because of these technological deficiencies, scientists were unable to ascertain the correct structure of the DNA bases – whether they existed in the keto or enol form, or whether they occurred in the amino or imino tautomer. Subsequently, Miriam's error went undetected for many years.

    Miriam's pH studies also had the unforeseen effect of contributing to the general sense of uncertainty regarding the structure of the DNA bases. During the 1940s and '50s, there was confusion regarding the structures of the DNA bases when they occurred in different physical states. Some believed that one tautomeric form existed when the DNA base was in a solid state, whereas another form occurred when the same DNA base was in a liquid solution. It was believed that one could not apply the knowledge gained from the x-ray diffraction studies of the solid crystals to the solutions of the DNA bases, where Miriam's worked suggested enolization was occurring. But even if one did make the assumption that the structures of the DNA bases were the same in both physical states, the x-ray diffraction studies at that time were unable to provide definite answers because they could not distinguish between the different tautomeric forms.[18] Also, the general uncertainty in the structure of the DNA bases found its expression in the textbooks of the day. One book would depict the enol form, while another would show the keto. Yet another would show the imino tautomer, whereas others would depict the the amino form. According to science historian Robert Olby, "Until this confusion was sorted out, Watson and Crick were doomed to fail in their quest for the structure of DNA."[19]

    The uncertainty regarding the structure of the DNA bases adversely affected early attempts to elucidate the structure of DNA. As described in chapter 4, a number of scientists were contemplating a helical shape for DNA, in part, because of Linus Pauling's discovery of the alpha helix for proteins. However, what was unclear at the time was the number of DNA strands and how those strands were bonded together. Faced with the mobile hydrogen atoms of the tautomeric bases, scientists were unable to conceive of how hydrogen bonding could occur between the DNA bases. Francis Crick recalled, "Where I made a complete mistake was assuming that each base actually existed in more than a single tautomeric form. Even if the less correct tautomer of each base only existed for 5 or 10 percent of the time this would make it difficult, I thought, to use the hydrogen bonds of the bases to build a regular structure. I therefore concluded, quite incorrectly, that hydrogen bonds could play little or no part in forming the structure."[20] Similar thoughts were echoed by Watson, who wrote, "Though for over a year Francis and I had dismissed the possibility that the bases formed regular hydrogen bonds, it was now obvious to me that we had done so incorrectly. The observation that one or more hydrogen atoms on each of the bases could move from one location to another (a tautomeric shift) had initially led us to conclude that all the possible tautomeric forms of a given base occurred in equal frequencies."[21] So instead of focusing on the DNA bases, Watson and Crick turned their attention to the phosphate groups as the force binding the DNA strands together.

Misled by tautomerism, the early model builders believed the phosphate groups were the bonding force holding DNA together. As a consequence, models were constructed with the sugar-phosphate backbones oriented toward the center of the helix, while the DNA bases faced outwards. For example, in 1951, before their successful double helix model, Watson and Crick initially built a triple helix model consisting of three, intertwined, sugar-phosphate strands held together by the attraction between the negatively-charged phosphates and positively-charged sodium ions. Similarly, in 1953, Linus Pauling and Robert Corey authored a paper describing a triple helix held together by hydrogen bonding between the phosphate groups.[22] However, after a short inspection of Watson and Crick's triple helix, Rosalind Franklin was quick to point out that DNA could absorb a lot of water, suggesting that the phosphates were surrounded by water and were unavailable as a force to hold the helix together. Franklin further suggested that the phosphates were positioned on the outside of the DNA molecule, where they could be easily hydrated. Consequently, Watson and Crick were forced to abandon their triple helix model of DNA.

Having decided to build two-stranded instead of three-stranded models because "important biological objects come in pairs,"[23] Watson and Crick returned to the issue of how to bind the DNA strands together. Looking for another force to hold the helix together, their attention turned to the DNA bases. Francis Crick recalled the occasion. "'Why not,' I said to Jim one evening, 'build models with the phosphates on the outside?' 'Because,' he said, 'that would be too easy' (meaning that there were too many models he could build in this way). 'Then why not try it?' I said, as Jim went up the steps into the night. Meaning that so far we had not been able to build even one satisfactory model, so that even one acceptable model would be an advance, even if it turned out not to be unique."[24] Watson had a similar remembrance. "Though I kept insisting that we should keep the [sugar-phosphate] backbone in the center, I knew none of my reasons held water. Finally over coffee, I admitted that my reluctance to place the bases inside partially arose from the suspicion that it would be possible to build an infinite number of models of this type. Then we would have the impossible task of deciding whether one was right. But the real stumbling block was the bases. As long as they were outside, we did not have to consider them."[25]

The first step in building their new "backbone-out" model was to construct a helical, sugar-phosphate chain. Watson recalled, "There was no difficulty in twisting an externally situated backbone into a shape compatible with the X-ray evidence,"[26] data collected by Rosalind Franklin and Raymond Gosling. Although Watson and Crick initially built just one of the two DNA strands, they realized from Franklin and Gosling's x-ray studies that the second sugar-phosphate chain would be oriented in the opposite direction to the first. Pleased that their new model was on the right track, Crick excitedly encouraged Watson to continue on with its refinements. But, according to Robert Olby, "Watson was not enthusiastic; for the problem that now faced him was what to do with the bases."[27]

The year before, Watson and Crick had met with Erwin Chargaff and learned of the DNA base ratios (adenine = thymine, guanine = cytosine). Chargaff had the following recollection of the two, curious "pitchmen."[28] "The first impression was indeed far from favorable; and it was not improved by the many farcical elements that enlivened the ensuing conversation,... So far as I could make out, they wanted, unencumbered by any knowledge of the chemistry involved, to fit DNA into a helix. The main reason seemed to be Pauling's alpha-helix model of a protein. I do not remember whether I was actually shown their scale model of a polynucleotide chain, but I do not believe so, since they still were unfamiliar with the chemical structure of the nucleotides. They were, however, extremely worried about the correct 'pitch' of their helix... I told them all I knew. If they had heard before about the pairing rules, they concealed it. But as they did not seem to know much about anything, I was not unduly surprised. I mentioned our early attempts to explain the complementarity relationships by the assumption that, in the nucleic acid chain, adenylic [adenine] was always next to thymidylic acid [thymine] and cytidylic [cytosine] next to guanylic acid [guanine]... It was clear to me that I was faced with a novelty: enormous ambition and aggressiveness, coupled with an almost complete ignorance of, and a contempt for, chemistry,..."[29] At the end of their meeting, Watson was left with as much regard for Chargaff as Chargaff had for him, and Watson was determined to solve the structure of DNA without using Chargaff's ratios. Crick, on the other hand, was excited by the meeting and went to the library to read Chargaff's papers.[30]

After completing the sugar-phosphate chain on their model, Watson and Crick then addressed the question of how to evenly pack irregularly-shaped DNA bases within their helix. First, Watson copied the structures of the DNA bases from J. N. Davidson's monograph *The*

Biochemistry of Nucleic Acid,[31] and then he attempted to find a solution by sketching them onto paper. However, unknown to Watson at that time, the DNA bases were incorrectly depicted in the book as enol tautomers rather than as keto structures. Of the four DNA bases, only adenine was pictured correctly. Watson recalled his frustration. "My aim was somehow to arrange the centrally located bases in such a way that the backbones on the outside were completely regular – that is, giving the sugar-phosphate groups of each nucleotide identical three-dimensional configurations. But each time I tried to come up with a solution I ran into the obstacle that the four bases each had a quite different shape. Moreover, there were many reasons to believe that the sequences of the bases of a given polynucleotide chain were very irregular. Thus, unless some very special trick existed, randomly twisting two polynucleotide chains around one another should result in a mess. In some places the bigger bases must touch each other, while in other regions, where the smaller bases would lie opposite each other, there must exist a gap or else their backbone region must buckle in."[32] Discouraged, Watson ceased the paper exercise and returned to the issue of hydrogen bonding.

Pyrimidine   Pyrimidine   (Too far apart)

Purine   Purine   (Too close together)

Purine   Pyrimidine   (Just right)

Although they had previously dismissed hydrogen bonding between the DNA bases because of the mobile hydrogen atoms associated with the tautomers, Watson revisited the topic. He surveyed the scientific literature and came to an important realization. Watson wrote, "J. M. Gulland's and D. O. Jordan's papers[33] on the acid and base titration of DNA made me finally appreciate the strength of their conclusion that a large fraction, if not all, of the bases formed hydrogen bonds to other bases. Even more important, these hydrogen bonds were present at very low DNA concentrations, strongly hinting that the bonds linked together bases in the same molecule. There was in addition the X-ray crystallographic result that each pure base so far examined formed as many irregular hydrogen bonds as stereochemically possible. Thus, conceivably the crux of the matter was a rule governing hydrogen bonding between bases."[34]

During his survey of the literature, Watson also read June Broomhead's thesis, which contained a diagram of a section of an adenine hydrochloride crystal showing hydrogen bonding between neighboring adenine bases. He was then struck with the idea that similar DNA bases, like two adenines, could form hydrogen bonds between the DNA strands. Watson recalled his "like-with-like pairing" revelation. "Suddenly I realized the potentially profound implications of a DNA structure in which the adenine residue formed hydrogen bonds similar to those found in crystals of pure adenine. If DNA was like this, each adenine residue would form two hydrogen bonds to an adenine residue related to it by a 180-degree rotation. Most important, two symmetrical hydrogen bonds could also hold together pairs of guanine, cytosine, or thymine. I thus started wondering whether each DNA molecule consisted of two chains with identical base sequences held together by hydrogen bonds between pairs of identical bases. There was the complication, however, that such a structure could not have a regular backbone, since the purines (adenine and guanine) and the pyrimidines (thymine and cytosine) have different shapes. The resulting backbone would have to show minor in-and-out buckles depending upon whether pairs of purines or pyrimidines were in the center."[35]

Watson's like-with-like pairing scheme (adenine hydrogen bonded to adenine, thymine with thymine, and so on) turned out to be another idea misguided by tautomerism. Jerry Donohue, an American crystallographer who shared the office with Watson and Crick, examined Watson's pairing scheme and told him that it was not feasible. Watson recalled, "The tautomeric forms I had copied out of Davidson's book were, in Jerry's opinion, incorrectly assigned. My immediate retort that

several other texts also pictured guanine and thymine in the enol form
cut no ice with Jerry. Happily he let out that for years organic chemists
had been arbitrarily favoring particular tautomeric forms over their al-
ternatives on only the flimsiest of grounds. In fact, organic-chemistry
textbooks were littered with pictures of highly improbable tautomeric
forms. The guanine picture I was thrusting toward his face was almost
certainly bogus. All his chemical intuition told him that it would occur
in the keto form. He was just as sure that thymine was also wrongly
assigned an enol configuration. Again he strongly favored the keto alternative... Jerry, however, did not give a foolproof reason for
preferring the keto forms."[36]

Donohue recalled discussing the tautomeric forms of the DNA
bases with Watson "more than once,"[37] and eventually Watson relented
and began model building with the keto structures. "Thoroughly worried, I went back to my desk hoping that some gimmick might emerge
to salvage the like-with-like idea," Watson remembered, "But it was
obvious that the new assignments were its death blow. Shifting the hydrogen atoms to their keto locations made the size differences between
the purines and pyrimidines even more important than would be the
case if the enol forms existed. Only by the most special pleading could
I imagine the polynucleotide backbone bending enough to accommodate irregular base sequences. Even this possibility vanished when
Francis came in. He immediately realized that a like-with-like structure
would give a 34 angstrom crystallographic repeat only if each chain
had a complete rotation every 68 angstroms. But this would mean that
the rotation angle between successive bases would be only 18 degrees,
a value Francis believed was absolutely ruled out by his recent fiddling
with the model. Also Francis did not like the fact that the structure gave
no explanation for the Chargaff rules [adenine = thymine, guanine =
cytosine]."[38]

Abandoning the like-with-like scheme, Watson then tried pairing
different combinations of the DNA bases to find those that could form
hydrogen bonds. By shuffling around cardboard cut-outs of the keto-
amino forms of the DNA bases, he soon found the pairing relationship
he was looking for. According to Watson, "When I got to our still
empty office the following morning, I quickly cleared away the papers
from my desk top so that I would have a large, flat surface on which to
form pairs of bases held together by hydrogen bonds. Though I initially
went back to my like-with-like prejudices, I saw all too well that they
led nowhere. When Jerry [Donohue] came in I looked up, saw that it
was not Francis, and began shifting the bases in and out of various

other pairing possibilities. Suddenly I became aware that an adenine-thymine pair held together by two hydrogen bonds was identical in shape to a guanine-cytosine pair held together by at least two hydrogen bonds. All the hydrogen bonds seemed to form naturally; no fudging was required to make the two types of base pairs identical in shape. Quickly I called Jerry over to ask him whether this time he had any objection to my new base pairs. When he said, 'no,' my morale skyrocketed, for I suspected that we now had the answer to the riddle of why the number of purine residues exactly equaled the number of pyrimidine residues. Two irregular sequence of bases could be regularly packed in the center of a helix if a purine always hydrogen-bonded to a pyrimidine. Furthermore, the hydrogen-bond requirement meant that adenine would always pair with thymine, while guanine could pair only with cytosine. Chargaff's rules then suddenly stood out as a consequence of the double-helical structure for DNA."[39]

The next question was whether the DNA base-pairs (adenine with thymine, guanine with cytosine) would fit within the sugar-phosphate backbone model devised earlier. By using metal pieces to represent the DNA bases and paying special attention to the position of the atoms relative to one another, Watson and Crick gradually assembled their double helix model. When completed, they had successfully stacked the pairs of DNA bases in the center of the helix. Crick summarized, "our model captured all the essential aspects of the double helix. The two helical chains, running antiparallel [in opposite directions], a feature I had deduced from Rosalind's own data; the backbone on the outside, with the bases stacked on the inside; and, above all, the key feature of the structure, the specific pairing of the bases."[40]

In the spring of 1953, Watson and Crick's paper[41] describing their double helix model of DNA was published along with articles authored by Maurice Wilkins[42] and Rosalind Franklin.[43] Although the proposed double helix model was consistent with Chargaff's ratios and Franklin's x-ray diffraction data, Watson and Crick's paper did not prove that DNA was a double helix. News of the double helix model was greeted with mixed responses, mostly skepticism. Chargaff, for example, wrote in 1955 that Watson and Crick's model had several attractive features, "But between what is plausible and what is true in the natural sciences there lies an ocean which has seldom been crossed."[44] Likewise, some believed that the double helix model would reach a similar demise as Pauling and Corey's triple helix suggestion. On the other hand, the double helix model of DNA was well received by other scientists, including Linus Pauling. Although Pauling and others felt that the

Watson and Crick structure was essentially correct, they understood that it was a provisional model subject to refinement. "Meanwhile," science historian Horace Judson summarized, "it was left to others to clean up the details of the Watson-Crick model of DNA and resolve some doubts that it inspired."[45]

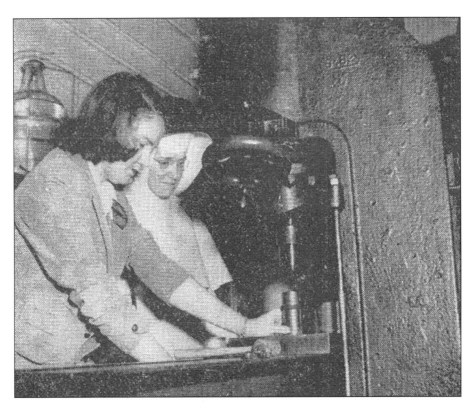

Sister Miriam Michael Stimson, OP, and a student
prepare a KBr disk using a press and die.

Sister Miriam Michael Stimson, OP,
her mother (Mary Stimson), and Pope Pius XII.

Chapter 9

KBr

*I attended many professional meetings with Sister Miriam. Many of
these were sponsored by the Spectroscopy Society of Pittsburgh.
Miriam was well known and recognized in the Society for her development of the KBr pellet technique used to study the IR spectrum of solid
compounds. She was affectionately introduced as Sister Stimson KBr.*
                                          - Sister Irma Gerber, OP

While Watson and Crick's double helix model of DNA slowly
gained acceptance within the scientific community, questions remained concerning the DNA base-pair combinations.
Watson and Crick had assumed that the DNA bases were in the keto
and amino tautomeric forms, and as such, had predicted that each adenine and thymine pair as well as every cytosine and guanine pair was
capable of forming two hydrogen bonds. However, some questioned
this assumption since, as Donohue summarized in 1956, "Precise X-ray
crystallographic information on these and related compounds is mea-
ger."[1] Recognizing the uncertainty in the structure determinations of
the purines and pyrimidines of DNA, Pauling and Corey reviewed the
existing x-ray diffraction data of the DNA bases that year and correctly
surmised that cytosine and guanine were capable of forming three hydrogen bonds, rather than the two suggested by Watson and Crick.
Pauling and Corey further cited evidence in support of Watson and
Crick's assumption regarding the tautomeric forms, but also wrote, "at
the present time only tentative conclusions can be reached"[2] because of
the lack of reliable x-ray data. This uncertainty in the tautomeric forms
even led some to propose alternative DNA base-pair combinations. For

example, a different guanine and cytosine pairing was suggested that involved the imino form of cytosine, rather than the amino tautomer.[3] While the method of x-ray diffraction failed to yield definitive answers in the late 1950s and early '60s, some scientists turned their attention to using a relatively new method called infrared spectroscopy. One such scientist was Sister Miriam Michael Stimson, OP.

Although it took several years for Miriam to realize the mistake with tautomerism, she always recognized the value of continuing her education after graduate school. For a number of years, she traveled to other institutions and conducted research over the summer. At Siena Heights College, she was one of only two scientists working at the College – teaching everything from chemistry to biology and physics. There was no breadth of expertise there, since both she and Sister Agnita Reuter were working on the same research project. In essence, she was working in isolation from the rest of the scientific community, unlike what she had experienced as a graduate student at the Institutum Divi Thomae. In the early 1940s, Miriam visited the Institutum to renew her contacts there. However, shortly after Miriam graduated from the Institutum Divi Thomae, Loofbourow left Cincinnati in 1940 and took a position in the spectroscopy laboratory at the Massachusetts Institute of Technology. In the meantime, Sperti's interest in wartime research projects diverted the Institutum's focus away from cancer and the biodynes, a direction that the Institutum would never regain.[4] Possibly as a result of the changing climate at the Institutum Divi Thomae, Miriam visited the Institutum less often and instead spent several summers conducting research at regional public universities, including Michigan State University and the University of Michigan.

While the Institutum Divi Thomae and Siena Heights College were welcoming women in science, Miriam encountered sexism while at the University of Michigan. She recalled, "When I went to the University of Michigan, the number of women allowed to enter the graduate school was limited to the number of ten, and we had to use the servants' entrance to the Union. I wasn't allowed to use the front door. But men could go into the League, which was the women's place."[5] "I remember when I went to U of M, the head of the chemistry department [Dr. Anderson] didn't want me. He didn't want women... He didn't believe women should be educated... In the '40s that was the way it was. So I found there was one lab at U of M, Professor [Kasimir] Fajans' lab that would allow women to work in the lab. There were about 5 or 6 of us that worked in that lab."[6]

Women in chemistry were – and still are – very much a minority, an observation that did not escape the attention of some of Miriam's students. Dorothy Nyhoff (O'Kane) commented, "Sister took us to Detroit once for some conference. That was the first time I started to realize that I had stepped into a male dominated field."[7] Another student recalled the following: "Women in science were still a real novelty back then."[8] "In 1956, she [Miriam] sent her three sophomore students to Tri-State College in Angola, Indiana for an ACS [American Chemical Society] student affiliate meeting. We three women were a very small minority among the mostly male engineering students, but we definitely held our own. I think she wanted us to see exactly what it was like to be considered an oddity."[9]

As a woman in chemistry, Miriam sometimes had to endure the patriarchal attitude of the male scientists, some of whom regarded "women as mere diversions from serious thinking."[10] "I was very much aware of the idea that women were downgraded in such fields as analytical chemistry," noted Miriam.[11] In those days, women were either prohibited from attending many public universities or their numbers were restricted, and those that entered the workforce often received less pay than their male counterparts. Furthermore, most female academicians were relegated to the low-paying, teaching roles at small women's colleges, rather than the higher paying, research positions at the large public institutions. Within this discriminatory environment, Miriam sometimes found that her ideas were either dismissed or disparaged by her male counterparts. A former student recounted the following anecdote. "Sister delighted in telling the story of her attendance at an important national meeting in her field – spectroscopy. She was at a dinner for the participants when a heated discussion occurred regarding a specific problem one of the men was having with his research project. When she attempted to offer a suggestion, he told her rudely that the only authority on the subject he would consider was M. Stimson. Unless you knew her when she was young and formidable, you can't even begin to imagine the expression of perverse delight on her face as she declared, 'I am M. Stimson!'"[12]

Miriam also believed she was treated differently from the other scientists not only because she was a woman, but because she was a religious sister. She explained, "After the Second World War when they had the National Science Foundation and had NSF workshops in the summer, I decided there were three classes of people: men, sisters, and women. There were very, very few laywomen. Eighty percent of the women who would attend would be nuns of different congregations, which I think is very interesting. I think universities, which admitted women earlier, saw sisters as different."[13] "Sister often remarked that being a woman scientist was difficult enough, but being a nun as well made the situation even harder," added a former student.[14] In those days, it was only acceptable for an unmarried woman to hold a job to support herself. Married women, on the other hand, were expected to give up their careers to bear and raise children. Before the women's movement of the 1960s, Catholic sisters, like Miriam, were among the first American feminists, advocating equal opportunities for women in education and the workplace. These women scientists encountered many obstacles from their male counterparts, some of whom felt that "the best home for a feminist was in another person's lab."[15] Some of Miriam's former students recalled that she was often brutally frank when describing the prejudices women faced in the public universities and the workforce, yet she also offered them words of encouragement. Miriam advised one young woman, "There are many opportunities ... by which a girl may use her interests and talents in the service of God... Every girl has the obligation to develop her God-given talents and to use them for the honor and glory of God whether she chooses the religious or the secular way of life."[16]

In addition to conducting research at the University of Michigan, Miriam also spent a summer at the Illinois Institute of Technology where she learned infrared spectroscopy. Infrared spectroscopy is a method of characterizing and identifying chemical compounds based on their absorption of infrared radiation, wavelengths of light lying just beyond the red limit of the visible spectrum. By this method, samples are exposed to a source of infrared radiation, and the wavelengths of infrared that are transmitted by the compound are then detected. The result, called an infrared spectrum, shows the percentages of infrared light that are absorbed by the compound at each wavelength, which are related to the particular atoms. Since the type and location of the atoms of different compounds will vary, each compound will have a unique infrared absorption spectrum, which can be used, like a fingerprint, as a source for identification.

Miriam later described this principle using the following analogy. By testing the ability of a compound to absorb different wavelengths of infrared light, scientists can gain a better understanding of the chemical structure of the compound. Likewise, "the situation of the chemical might be compared to that of a man confronted with a series of preference items: soups (bean, chicken, onion, pea, tomato, vegetable), salad, etc. The set of questions tests his appetite for foods, … and we can eventually form some impression as to the type of a man."[17]

In the early days of infrared spectroscopy, Miriam and the other spectroscopists faced many experimental challenges. The early infrared pioneers not only had to design and build their own instruments, but the individual components too, including the prism, mirrors, and radiometer. Obtaining an infrared spectrum was also a tedious process. Since each point on the spectrum had to be measured individually, recording an infrared spectrum often required three to four hours or more.[18] At the Illinois Institute of Technology, Miriam worked in the spectroscopy laboratory, which was housed in the basement of an 1880-vintage building standing between the Rock Island Railroad tracks and Lake Michigan.[19] Infrared spectrophotometers were usually located on the basement floor and the measurements were often made at night to minimize the effect of vibrations.[20] Yet it was there, with an instrument made in the local shop, using galvanometer tracings made in spite of the vibrations of too-frequent trains, that Miriam learned the rudiments of infrared spectroscopy.[21]

In many respects, infrared spectroscopy is a more powerful method than UV spectroscopy for characterizing chemical compounds. Miriam's former mentor John Loofbourow explained, "For many biochemical problems, infrared absorption spectra offer aid that cannot be obtained by any other physical methods, particularly ultraviolet spectra. A wide variety of highly important biological substances – hormones, amino acids, carbohydrates, and saturated fatty substances, to cite several classes – exhibit either little ultraviolet absorption at wavelengths above 2000 A [angstroms] or else show absorption characteristic only of a small chromophoric group that does not enable the biochemist to identify or analyze for the remainder of the molecule. On the other hand, the infrared spectra of these substances, besides revealing the presence of specific groups, exhibit features characteristic of the molecule as a whole. For example, infrared examination of two compounds that differ only in being geometrical isomers nearly always makes possible a sharp distinction between them."[22]

Realizing the limitations of using UV spectroscopy to study the structure of the DNA bases, Miriam turned her attention to using infrared spectroscopy. In 1948, with the help of a 5 year grant from the Research Corporation of New York, Miriam and Siena Heights College were able to purchase a Beckman model IR-2 spectrophotometer for infrared spectroscopy. The Beckman model IR-2 was one of the first infrared spectrophotometers to fully incorporate vacuum tube electronics. The IR-2 was a powerful instrument for its day, but its successful operation required closely controlled environments, relatively long periods of time, and the skill of a highly trained spectroscopist.[23] Using one of the most advanced instruments of that time, Miriam planned to address the uncertainty regarding the structures of the purines and pyrimidines of DNA.

When Miriam began using infrared spectroscopy, there were problems associated with the sampling techniques used to obtain an infrared spectrum of many compounds, including the nucleic acids and the DNA bases. Samples were commonly prepared for infrared spectroscopy in either of two ways; they were either dissolved with a solvent to form a liquid solution or the dried samples were mixed with an oil to form a mull. Both of these methods were problematic, because the solvents and oils absorbed infrared light and obscured the absorption by the sample compound. Also, many biologically important molecules are not soluble in the solvents commonly used for infrared spectroscopy. Furthermore, it was difficult to obtain quantitative results with oil mulls, due to the difficulties in controlling sample concentration and film thickness.[24] "Thus, before information available from infrared spectrophotometry could be applied to many compounds of biological or pharmaceutical interest, methods for sampling of complex solids had to be developed," summarized Miriam.[25]

Miriam addressed the sampling problem with the assistance of Sister Marie Joannes O'Donnell, OP, an Adrian Dominican and former student. Sister Marie Joannes joined the faculty at Siena Heights College in 1948 when she replaced Sister Agnita Reuter, who left to become principal of Sacred Heart Academy in Puerto Rico.[26] Sister Marie Joannes recalled that both she and Miriam "had essentially the same background, which was chemistry." "Back in those days you got appointed to teach all the sciences, you taught all the sciences. So, we split up the biology, and I taught 4 or 5 courses, not all at the same time, and she did all the rest of them, and I did the physics."[27] Sister Marie Joannes was a popular teacher, and one of her former students remembered her as "a very lovely person, extremely well-liked by the students."[28] Outside the classroom, the two Sisters tackled the problem of preparing samples for infrared spectroscopy.

As a first step in addressing the sampling problem, Miriam reviewed the scientific literature. At that time, infrared samples were commonly studied in liquids. However, in 1948, Loofbourow and others at MIT suggested "that if a solid can be fabricated by rolling, melting, deposition, and similar methods into a thin sheet or plate of proper thickness then the solid itself could be examined."[29] Two years later, Loofbourow and his colleagues employed a vacuum technique whereby it became possible to prepare solid samples for study by ultraviolet spectroscopy.[30] In the meantime, Cook and others at the Institutum Divi Thomae were working on a procedure to examine solid protein samples in the infrared region of the spectrum by mixing the samples with potassium bromide (KBr).[31] These studies then led Miriam to the idea of developing a pressed disk technique using KBr as a nonabsorbing diluent.

In short, the KBr disk technique that the two Adrian Dominican sisters developed involved using a mortar and pestle to mix a finely divided sample intimately with KBr powder and then pressing the mixture in a die to form a translucent disk.[32] KBr was selected as the matrix material for a number of technical reasons. Unlike many solvents, KBr is transparent to infrared light, so it will not mask the absorbance of the sample compound. In addition, KBr has a relatively low sintering pressure, so under moderate pressures KBr will easily fuse with the sample compound.[33] "I tried other things," commented Miriam, "but I couldn't fuse them."[34] Furthermore, KBr was readily available in pure form. Finding mortars and pestles to grind and mix the KBr with the sample was relatively easy, since both are common laboratory supplies. However, obtaining the necessary die and press was another matter.

Miriam needed a die, a mold within which the sample and the KBr could be pressed together to form a thin, glass-like disk. At the time, there was not a die that could modified for this task, so a die had to be specifically created for this use. One of Miriam's students had a father who worked for the Chrysler Corporation, the automobile manufacturer, and through this connection, Miriam was able to get Chrysler to make a set of dies. "So, I went to Chrysler, and I had a sketch of the kind of die I wanted," described Miriam. "At Chrysler, they gave me a choice of through-hardened or case-hardened steel. I knew what I wanted, so I got case-hardened steel... I'll tell you one of the things that impressed me there. I don't know how they did any research, because everything was absolutely shiny. In other words, the one test tube that the chemist was using was in his hand. There was nothing else around!"[35]

The finished die consisted of two main pieces. The bottom piece was called the "anvil," upon which the sample and KBr mixture were carefully placed. Above the anvil, was the "plunger," which when lowered, would contact the anvil along "absolutely smooth"[36] faces and press the sample and KBr into a disk. Each KBr disk was about 1 millimeter thick and "just a little bigger than an aspirin."[37] The anvil and plunger pieces were connected by a holding ring, and the assembled die was enclosed in a steel cylinder, which was attached to a vacuum pump to remove air from the sample.

Next, Miriam needed a hydraulic press to form the KBr disks. Since she did not have a press in her laboratory, she initially tried using one located in a tool and die shop of a local factory during the men's lunch hour.[38] Sister Marie Joannes recalled, "We went in town ... to the tool and die place to use the hydraulic press. We went down with these big tall gauges because we had to do [measurements of] the pressure, and the pump, and evacuated them. ... and put them under the big thing [press]. This was to the entertainment of the gentlemen there. We were in these long, white habits, and we were in among the men in the tool and die department with their machines. We took a cab down and carried all our stuff with us. They did absolutely everything to make things more convenient for us."[39] Despite the kind offerings, the limited hours that the press was available, the need to transport equipment back and forth, and dust formed by normal factory work hampered their progress.[40]

An invitation to speak before the Adrian Technical Club, a small group of local businessmen interested in scientific research, indirectly answered Miriam's need for a press.[41] After her talk, there followed an excited two-hour discussion, and one of the businessmen called Miriam the next day asking if she had the time to look at a press that was currently not in use.[42] The unused 20-ton hydraulic press had been confiscated by the U.S. Air Force after World War II from Germany, where it was used to extrude propellers for airplanes.[43] Miriam found the press satisfactory, so the Adrian Technical Club arranged for the loan and transport of the war surplus machinery to Miriam's laboratory, where it was converted for scientific research. The press was "absolutely huge," recalled Sister Marie Joannes, "It really took up the whole outer room [of the laboratory]."[44] Diana Albera Luciani, a former student, had a similar memory. "I recall her lab as being two rather small rooms, an office with instruments, cabinets full of chemicals and a worktable, and a room dominated by a giant press... Her lab smelled like a medical clinic."[45] After the press was installed, it was subsequently tested. Sister Marie Joannes remembered the first test run. "The worker wanted to try it to see if it would work, and he hit the die, which is a cylinder, at an angle, and it shot out across the room ... broke the window!"[46]

The size, tremendous noise, and power of the hydraulic press initially made its operation a frightening experience. To make the KBr disks, the die was held absolutely still by hand while the press came down and applied "20,000 pounds per square inch pressure."[47] According to Miriam, "That [press] was a very fearsome thing, and it was hard on me. You had these steel chambers that you evacuated. In the middle of that were the dies you were using, and they had to be precisely seated or you would have disaster. And then this press we had gotten, the German press. A VERY big thing. Put your hand close enough… and have it coming down… to hold it just so… I was under so much stress. I taught students how to do this. Don't think that I wasn't a nervous wreck that their fingers might be in the way!"[48]

Employing this press, the two sisters approached the solution to the sampling problem. By using an empirical approach, they slowly worked out the optimal conditions for preparing the KBr disks. Other than deciding upon KBr as the matrix material, the two sisters also had to determine the best temperature and drying time for the KBr, the sample concentrations, the proportions of sample and KBr to use, the mixing time, the duration of sample evacuation, and how long the press should be applied, which according to a former student was the length of time required to pray two *Hail Marys*.[49] Developing the KBr technique was a tedious process, since each time they tested a variable, they would have to prepare a new disk. "I didn't do it overnight," commented Miriam.[50] This process was further complicated by the need to obtain an infrared spectrum in order to assess each trial disk.

Obtaining an infrared spectrum was a laborious process. Sister Marie Joannes recalled the time-consuming procedure. "In our time, we did [an infrared spectrum] point-by-point. We graphed it by hand. It took all afternoon. It took about three hours to get data for the section of the IR spectrum that we wanted… to get all of the little 'wiggles' of an IR spectrum. Your moving needle [of the IR spectrophotometer] doesn't give you a reading – it just gives you a zero point. So you have to move the dials until you get a null the way you would with a galvanometer. You have to get every shoulder as well as every maximum. Every little thing that could be done very quickly now was laborious. Just from pure expenditure of time and energy, it was a very full process, and the fact that made it worthwhile was that it [the KBr disk technique] was such an excellent new method."[51]

Eventually, the two Adrian Dominican sisters worked out the conditions for the KBr disk technique, and they were able to record the infrared spectra of guanine and cytosine. As they had predicted, their new procedure allowed for the study of the same physical sample in both the infrared and ultraviolet regions of the spectrum. Furthermore, examination of the resulting spectra revealed that the KBr disk technique was a significant improvement over the use of oil mulls. With the KBr disk method, there was an absence of interfering bands, lower scattering losses, higher resolution of spectra, better control of concentration and homogeneity of sample, ease in examining small samples, and possibility of storage of specimens for further studies.[52] Yet, the largest single advantage of the KBr disk technique over the oil mull was the ability to carry out quantitative, rather than qualitative, studies.[53]

In the summer of 1951, Miriam presented the new KBr procedure to the scientists attending the Symposium on Molecular Structure and Spectroscopy at The Ohio State University. In her abstract entitled, "The Preparation of Samples in the Solid State for Spectroscopic Study in the Infrared and Ultraviolet Regions," she wrote the following: "The desirability of correlating both the ultraviolet and infrared absorption for purposes of characterization and for theoretical consideration is recognized. However, the limitation imposed by the solubility of many organic solids makes uniform conditions for solution work over this range of the spectrum impossible. The further problems of solvent-sample interaction and of solvent absorption, where this is selective, cannot be overlooked. A method for readily preparing organic materials for spectrophotometric study in both the ultraviolet and infrared regions has been developed. This method involves the thorough mixing of the finely ground organic with solid potassium bromide and subsequent pressing into disks. The method to be described has been employed for study of the free base of guanine and also for the hydrochloride, and also has been used for anhydrous cytosine and for the crystalline hydrate. Other types of compounds also will be shown. The range of concentrations and disk thicknesses which will permit quantitative determination of intensity as expressed by molecular extinction coefficients will be discussed."

Miriam's talk was enthusiastically received by the scientific community, who immediately recognized the breakthrough that the KBr disk technique represented. This new sampling procedure, which circumvented the absorption and solubility problems associated with the use of solvents and oil mulls, was a significant advance in the quantitative analysis of biological and inorganic compounds. Furthermore, since proper sample preparation is paramount to obtaining an infrared spectrum, the KBr disk technique, along with the introduction of commercially available scanning and recording spectrophotometers, helped to make infrared spectroscopy one of the most powerful and widespread methods of chemical analysis in the 1950s and '60s. In addition, the KBr disk technique, which was likened to paper chromatography, was a relatively simple procedure with a plethora of applications, ranging from pharmaceutical drugs to industrial paints. Furthermore, the procedure was as useful in basic research as well as in applied work. Realizing the importance of Miriam's discovery, requests began pouring in from laboratories all around the country, as well as from Europe, for more details and reprints of her talk.[54] That summer, the two sisters then completed a manuscript describing their new procedure, which was published the following spring in the *Journal of the American Chemical Society*.[55]

In recognition of her achievement, Miriam was subsequently honored with invitations to speak at numerous conferences. In the fall of 1951, for instance, she presented the results of their work at the annual conference of the Institutum Divi Thomae, her alma mater. Speaking to this assembly of cancer experts, Miriam discussed the application of quantitative infrared spectrophotometry to the study of cellular metabolism.[56] Then came the formal invitation to deliver the endowed Peter C. Reilly Lecture in Chemistry at the University of Notre Dame, the first woman so honored.[57] After that, she was invited to lecture at the International Reunion for Molecular Spectroscopy at the Sorbonne in Paris, where she was honored as the first woman to speak since Nobel laureate Madame Marie Curie. Miriam recalled, "When I got the invitation to talk at the Sorbonne, Sister Gonzaga and I shared a suite of rooms with a bath between us. Anyhow, I was telling 'Gonzie' that I had this invitation... If we don't have the money, then I won't be able to go. So I wasn't saying anything about it. But 'Gonzie' talked to Mother Gerald. So then, Mother Gerald sends for me and wants to know why I didn't come to her and tell her about this. In those days, whenever sisters went anywhere there were always two. Well, Mother Gerald wanted me to go to the Sorbonne, but she didn't want to pay. I don't know how she did this, but she convinced my father to pay the way to go, and she said your mother can be your companion. So my father paid for my mother and me."[58]

Miriam's trip to Europe, her first journey overseas, was one of the highlights of her life – a fitting reward for her scientific accomplishments, years of dedication to her faith-inspired research, and perseverance in the face of personal and professional adversity. Miriam was accompanied by her mother, who traveled with her for about a month through France, England, Germany, and Italy. Miriam's father also joined them for approximately one week in Germany. Unaccustomed to the motions on passenger ships, Miriam described her ocean voyage: "I'm one of these people that get very sea sick. On going over, I made friends with a doctor from Chicago and he kept giving me the right meds [medicines]. So I did pretty well." However, "Coming back, he wasn't on there!"[59]

By the time she reached the lecture hall at the Sorbonne in Paris, Miriam was an experienced speaker – having given the KBr presentation before as well as having lectured chemistry at Siena Heights College for over ten years. Although she was the only female speaker at the meeting, "I did not feel a bit intimidated," claimed Miriam, who by this time was used to being treated as a novelty.[60] Although later she confided, "I was nervous" speaking before a foreign audience.[61] Regarding her talk, she recalled having problems with the lighting in the room. "When I got there, to the Sorbonne, they had this big sky light – wide open. And I thought, 'How am I going to show my slides?' Because in those days I had lantern slides, which were glass, and I wanted to show my spectra … on the screen. When I talked to LeConte, who was a professor in Paris that was putting this thing on, he was saying they had the covering off because it rained last week. But the point is, it wasn't raining that day. I talked to him [Quarianno] about this. And he went up, and he closed the shade over the skylight so I could show my graphs. When I got to the lectern, they had the lights off in the room, but they had no light on at the lectern. So, here I am trying to present a paper which I couldn't see!"[62] Afterwards, Miriam shared ideas with scientists from a dozen countries. During lunch, they drew pictures on the paper tablecloths to communicate subjects that would leave the most facile linguist groping for words.[63]

Following the week-long meeting in Paris, Miriam visited a number of research centers. For instance, she traveled to Bristol, where Dr. J. McOmie used the paper chromatography technique that she had just missed publishing.[64] In addition, she saw her KBr disks used in the Chester Beatty Institute of Cancer Research, London.[65] She also visited the Max Planck Institute in Tübingen, Germany. There, she met Dr. Ulrich Schiedt, whom she had met earlier at her Notre Dame talk. Schiedt, a German biochemist, had developed a similar KBr disk procedure around the same time as Miriam. With knowledge of Miriam's work and citing Miriam's abstract from the 1951 meeting at The Ohio State University as well as her 1952 *Journal of the American Chemical Society* KBr manuscript, Schiedt published his KBr technique in the

Zeitschrift für Naturforschung near the end of 1952.[66] Unlike Miriam, Schiedt applied the KBr disk technique to the study of amino acids, and he published another paper on the topic in 1953,[67] before his untimely death four years later. At the Max Planck Institute, Miriam admired the fine equipment in the spectroscopy lab and had to explain to bewildered German scientists, "But we don't have the Marshal Plan in Adrian [Michigan]."[68]

Miriam's journeys also included a spiritual visit to the birthplace of the Adrian Dominican Congregation. She went to the cloister of the Holy Cross in Regensburg, Bavaria, from which four sisters had ventured to the United States more than one hundred years before.[69] During her stay at the monastery, Miriam recalled that the German nuns had the custom of wearing brown shower gowns while privately bathing. Not accustomed to wearing anything while bathing, Miriam did not use one. Since Miriam's shower gown was always dry, the German nuns, "probably thought that I never bathed," remarked Miriam.[70]

Miriam's accomplishments were also recognized by the Papacy, and she was granted an audience with Pope Pius XII. Accompanied by her mother, she met the Pope at one of the summer estates outside of Rome, possibly Castel Gandolfo, rather than in Vatican City. "That was in the days before air conditioning, and Rome was notoriously warm," explained Miriam.[71] Despite the summer temperatures and humidity, those who visited the Pope at the Italian estate were expected to be conservatively dressed and fully clothed. Miriam and the other visiting Dominican sisters were dressed in full garb for the occasion, including their cloaks. On the other hand, Miriam's mother, unaware of the strict clothing requirements, mistakenly wore a short-sleeved dress that exposed her forearms. According to Miriam, this *faux pas* caused quite the stir with the papal officials, who wanted to cover her bare arms using Miriam's cloak. The ensuing commotion upset Miriam and had the effect of blocking out all memory of any inspirational words that the Pope may have uttered on that day.

After Miriam returned to her modest surroundings in Adrian, Michigan, she continued to receive much acclaim for the development of the KBr disk technique. In 1954, she published two more research papers, one in French, on the use of the new sample preparation technique in infrared spectroscopy.[72] The same year, the KBr disk technique was heralded as "The greatest recent improvement in the handling of samples ... In some laboratories this technique has supplanted all other solid phase methods."[73] Furthermore, a year later, the

KBr disk technique was praised as "a very real advance in infrared analytical technique" in a book co-edited by Erwin Chargaff.[74] Eventually, this new sampling procedure found its place within infrared spectroscopy textbooks[75] as well as numerous monographs.[76] Her KBr success also led to a number of small research grant awards from the American Cancer Society, which recognized the importance of funding basic research related to DNA. Miriam also received much attention from the Catholic and secular newspapers, appearing in over 40 articles between 1950 and 1960. Perhaps her most unusual tribute, though, came in 1959, when Miriam starred in a 36-minute-long educational movie entitled, "Infrared Spectroscopy." In the film, which premiered to the 5,000 scientists attending the Pittsburgh Conference on Analytical Chemistry and Applied Spectroscopy, the habit-clad scientist used a small press to form glass-like disks of KBr.

Despite the well-deserved notoriety, a close friend recalled that Miriam remained a humble and genuine woman.[77] Miriam's humility may have been fostered by a newspaper commentary authored by Elton Cook, her former research mentor from the Institutum Divi Thomae. In the article, Cook described a number of scientific advances (i.e., germ theory of disease, antiseptics, and the treatment of diseases), but then placed the scientific findings into a spiritual context. He wrote, "When the scientist looks at these accomplishments, made in the space of only a few years, he is proud, and rightly so. But it is easy to become over proud, to take too much credit, to feel that these results have been attained by men alone. It is easy to forget that, after all, we are merely realizing potentialities that have been present all along; that, in the truest sense of the word, we are not creating at all, but are really discovering the extent of God's wisdom. We are slowly learning how to live in this world. This, of course, does not in the least belittle the accomplishments of science. On the contrary, by regarding these accomplishments with humility, we magnify them and comprehend their true significance."[78] Miriam kept this article in a place separate from her other newspaper clippings, and judging by its worn appearance, she may have consulted it often.

Chapter 10

Unveiled

Jesus said, "Know what is before your face, and what is hidden from you will be disclosed to you. For there is nothing hidden that will not be revealed."

– The Gospel of Thomas

*T*he development of the KBr disk technique of preparing samples for infrared spectroscopy was an important step in resolving the controversy regarding the structure of the DNA bases. In addition to Miriam's published work on cytosine, others used the KBr disk method, along with oil mulls, to provide infrared evidence for the keto-amino forms of cytosine and guanine, the amino form of adenine, and the diketo form of uracil.[1] Furthermore, comparison of the infrared spectra of these compounds with those of the chemically synthesized methyl derivatives firmly established that the DNA bases existed in the keto and amino forms.[2] These results, along with the data obtained from ultraviolet spectroscopy, x-ray diffraction, and nuclear magnetic resonance studies confirmed the keto-amino form as the predominant tautomer of the DNA bases. Hence, Watson's guess that the DNA bases occurred in the keto and amino forms rather than the enol and imino configuration was correct, and in 1962, James Watson, Francis Crick, and Maurice Wilkins were awarded the Nobel Prize.

The discovery of the DNA double helix is considered by many to be the greatest scientific achievement of the twentieth century – rivaled by the discovery of the atom and the theory of evolution in the history of science. Like finding an essential puzzle piece that enabled other pieces to fall into place, the Watson and Crick model of DNA led to a multitude of other discoveries. The DNA double helix presented scientists with a workable framework from which to predict future discoveries. Much of how a gene functioned, for instance, could now be inferred indirectly from its structure. According to neurobiologist Gunther Stent, "It was to provide the highroad to understanding how biological information is encoded in the gene ..."[3] The DNA double helix also had the effect of redirecting the course of science. Prior to Watson and Crick's discovery, the physical sciences, the study of molecules through chemical means, was the predominant field of study. After the discovery, scientists turned their attention to understanding how such molecules in cells affected life, and so began a new era of cellular research called molecular biology.

Once the structure of DNA was known, a mechanism for its replication was immediately revealed. The model of DNA proposed by Watson and Crick consisted of two chains of nucleotides, where the two chains were held together by hydrogen bonding between complementary pairs of DNA bases. Since adenine always pairs with thymine and guanine with cytosine, if the base sequence of one chain was known, then the base sequence of its partner could easily be figured out. Conceptually, it was then easy to imagine how one chain could serve as a template for the synthesis of the second complementary chain, and vice versa. Realizing this, Watson and Crick wrote at the end of their first 1953 *Nature* paper the infamous understatement, "It has not escaped our notice that the specific pairing we have postulated immediately suggests a possible copying mechanism for the genetic material."[4] How they envisioned this process was later described in their second *Nature* paper: "Now our model for deoxyribonucleic acid is, in effect, a pair of templates, each of which is complementary to the other. We imagine that prior to duplication the hydrogen bonds are broken, and the two chains unwind and separate. Each chain then acts as a template for the formation on to itself of a new companion chain, so that eventually we shall have two pairs of chains, where we only had one before. Moreover, the sequence of the pairs of bases will have been duplicated exactly."[5] Watson and Crick's insight finally explained how genes could faithfully duplicate prior to transmission from one generation to the next, and it provided the physical explanation that was missing from Gregor Mendel's pioneering studies of heredity.

The DNA double helix also suggested, for the first time, a physical explanation for the appearance of mutations. A mutation refers to any permanent change from that which is normally present. Although the majority of mutations are deleterious, some can actually be beneficial.

For instance, mutations created the variations in seed texture, flower color, and stem length that Gregor Mendel observed with the garden pea. Since these mutations were heritable, in other words they could be passed on from one generation to the next, Mendel and others could study the transmission of these traits and develop the concept of a gene. Furthermore, mutations have allowed other scientists to create genetic maps, diagrams of the locations of genes on chromosomes, which in turn have assisted in the sequencing of entire genomes, all the DNA located within a normal cell. Mutations also play a critical role in evolution. By providing the genetic variation upon which natural selection can act, mutations form the basis for evolutionary change. Prior to the DNA double helix, though, scientists did not understand what mutations were at the cellular level.

Then in 1953, as a direct consequence of their structure of DNA, Watson and Crick offered the first physical explanation of mutation. In their second *Nature* paper, they reasoned that if the sequence of the DNA bases encoded information, then a change in that sequence could result in a mutation. They speculated, "Spontaneous mutation may be due to a base occasionally occurring in one of its less likely tautomeric forms."[6] For example, if the amino form of adenine were to change to the rare imino tautomer, then instead of hydrogen bonding with thymine, adenine could pair with cytosine. During DNA replication, this unusual pairing would then allow cytosine to become incorporated into a new DNA chain where thymine was expected, which would lead to a mutation if left uncorrected. Subsequent DNA molecules would then contain a guanine-cytosine base pair in place of the normal adenine-thymine combination. Watson and Crick's explanation, however, was not restricted solely to spontaneous anomalies caused by tautomerization, but instead suggested that all mutations were a change in the sequence of the DNA bases.

Although mutations can occur spontaneously at a low frequency, a much higher rate of mutation can be induced by various agents known as mutagens. Long before the DNA double helix was realized, scientists had known that "irritants," such as coal tar and radiation, could induce mutations, and in many cases, cancer too. Then in 1953, Watson and Crick's explanation of mutation opened up the possibility that the various mutagens, in one way or another, were all exerting their effects by altering the DNA base sequence. One of the initial tests of this idea utilized 5-bromouracil, an analog of thymine. Since 5-bromouracil is structurally similar to thymine, except that it contains bromine, organisms can substitute it for thymine in their DNA. However, this analog differs from thymine, which occurs in the keto form, in that 5-bromouracil contains a high proportion of the enol tautomer. When organisms, like bacteria, were treated with 5-bromouracil, they incorporated 5-bromouracil in their DNA in place of thymine. Then, instead of bonding with adenine, the enol form of 5-bromouracil paired with guanine, and as predicted, led to a DNA base substitution mutation. Subsequent research involving other mutagens, such as the coal tar derivative acridine and ultraviolet radiation, confirmed the finding that mutagens damage DNA by altering the DNA base sequence.

The value of using DNA base analogs as medicinal drugs was soon realized. In the 1950s, George Hitchings and Gertrude Elion, who themselves received the Nobel Prize in 1988, synthesized diaminopurine and thioguanine, analogs of adenine and guanine, respectively.[7] They understood that the DNA base analogs were incorporated only in the DNA of dividing cells during DNA replication. These cells would develop mutations that would inhibit their growth or cause their death. On the other hand, cells that are neither dividing nor making new DNA do not take up the DNA base analogs and escape from this type of mutation. Since cancerous tissues consist of rapidly dividing cells that undergo DNA replication, Hitchings and Elion decided to test the DNA base analogs as potential cancer drugs. When tested, the two DNA analogs proved to be effective treatments for leukemia, a form of cancer characterized by unusually high levels of white blood cells. Other DNA base analogs were also found to be effective in destroying cancer cells, and today these types of compounds form the basis for many chemotherapy treatments.

The DNA base analogs were also found to be effective drugs against certain viruses. Viruses are tiny particles of nucleic acid that are encapsulated within a coat of protein. Some viruses use DNA as their genetic material, while others, such as the human immunodeficiency virus (HIV) utilize RNA. After infection, HIV makes a DNA copy of its RNA. This DNA copy then invades the host's DNA, where it initiates the synthesis of new viruses, and eventually causes acquired immunodeficiency syndrome (AIDS). The best known AIDS drug is azidothymidine (AZT), an analog of thymine. Like other thymine analogs, AZT can substitute for thymine during DNA synthesis. However, instead of inducing mutations, such as those caused by 5-bromouracil, incorporation of AZT into viral DNA prevents further DNA synthesis. Without DNA, HIV is unable to replicate and cause disease.

Viruses have also been implicated with inducing certain types of cancers. Although Peyton Rous had shown in the early 1900s that an entity isolated from a cancerous chicken could elicit disease in a healthy chicken, the "infectious agent" hypothesis of cancer was largely ignored, because there was little evidence that cancer was contagious. But then in the 1960s, Rous' cancer-inducing entity was identified as a virus, which led to the discovery of other animal cancer viruses. For instance, the human papilloma virus was found to infect a woman's cervical cells and initiate the development of cervical cancer. With such knowledge, scientists were then able to successfully create and test a vaccine against the human papilloma virus – a monumental step towards the prevention of virus-initiated cancers.

Analysis of the Rous sarcoma virus and others revealed that they contained cancer-causing genes termed "oncogenes." When noncancerous cells were treated with these oncogenes, some formed cancerous growths – demonstrating that the genes alone could cause cancer. Later, it was discovered that the viral oncogenes were not of viral origin at all, but instead were versions of genes possessed by humans, chickens, and other animals. The oncogenes turned out to be mutated forms of normal genes, called "proto-oncogenes," which the virus had confiscated from its host. Proto-oncogenes initiate cell division so that we can grow from a fertilized egg into an adult and also heal our wounds. But it is important that these genes be active only when and where they are needed. If mutated by too much cigarette smoke or UV radiation, for instance, then these genes may become activated all of the time and induce uncontrolled cell division and tumor formation. Fortunately, our cells also possess genes that inhibit cell division. These genes are called "tumor suppressor genes."

Tumor suppressor genes prevent tumor formation by terminating normal cell division. For example, when you accidentally cut your finger, the cells immediately surrounding the wound begin to divide to make new cells. After a sufficient number of cells have been produced to close the wound, tumor suppressor genes direct the cells to stop dividing. By inhibiting cell division, tumor suppressor genes also help to prevent the formation of cancers. On the other hand, if the tumor suppressor genes themselves become mutated, perhaps by exposure to coal tar, for example, then the cell will be unable to stop proliferating, in which case a tumor will eventually develop. Interestingly, such research has yielded insight as to how chemotherapy and radiation therapy may work. Some scientists believe that in addition to directly killing some cancer cells, the chemicals and radiation used to treat cancers may activate certain tumor suppressor genes in cancerous cells, which then induce those cells to commit suicide. This idea has led to the clinical testing of the proteins of tumor suppressor genes as potential cancer drugs.

Discovery of the DNA double helix led to the unveiling of cancer, so to speak, because it revealed cancer as a genetic disease. With knowledge of the structure of DNA, scientists were able to understand the nature of mutations and ascertain the commonalities between "chronic irritants," like coal tar and x-rays, and "infectious agents," such as viruses. According to Watson, "If we had not been able to study cancer at the level of the change in DNA that starts it, the disease would still be a hopeless field."[8] Knowledge of the genetic basis of cancer has since revolutionized the way cancers are treated. Colon purges, magic crystals, and crow's feet have been supplanted by DNA testing, chemotherapy, radiation therapy, and a host of DNA-based drugs. Such treatments have provided cures for many and helped others to manage their cancers and to prolong their lives. Although a single cure-all for cancer remains elusive, the promise of finding effective treatments for each and every type of cancer is greater today than it has ever been.

Similar to the unveiling of cancer, Miriam too was unveiled in the 1960s. The 1960s was a tumultuous decade of change – sparked by the civil rights movement, the Vietnam War, and the women's movement. In those years, the Catholic Church and the Adrian Dominicans also underwent major reforms. During that period, the Adrian Dominican Sisters turned outwards to the world, and many of the sisters took back their baptismal names and exchanged their habits for lay clothing. Some of these changes were not welcomed by all, which sparked conflicts between the traditional, "old" Catholics and the reform-minded, "new" Catholics. During those years, Miriam was a vocal advocate for change. As a result of the reforms, Miriam found a new sense of freedom from the previously restrictive environment. Her black veil gradually gave way to her signature hats. She moved out of the residence hall, learned how to drive a car, and moved into a house. Miriam traveled to Russia and China, and she even acquired several pet animals. However, frustrated by the traditionalists in the congregation as well as at Siena Heights College, Miriam left Adrian in 1969. After a nine year hiatus, during which she taught at Keuka College and cared for her ailing parents, Miriam returned to Siena Heights College as the Director of Graduate Studies. After developing several successful graduate programs, Miriam retired in 1991.

During her retirement, Miriam faced a new and more treacherous adversary – cancer. She was diagnosed with stage II breast cancer; the cancerous cells had not yet spread to other parts of her body. Following her diagnosis, Miriam immediately began reading and learning as much as she could about her new foe, and typical of her analytical mind, she even calculated her survival rate.[9] Miriam underwent the standard cancer treatment regime, including chemotherapy; Derivatives of the DNA bases that she had studied for so long coursed through her veins. After all the treatments were completed, following much pain and suffering, Miriam, yet again, emerged victorious.

Miriam was not your stereotypical nun, nor was she a typical scientist. Many knew of her, but few really knew her as well as her former student, colleague, and friend, Sister Sharon Weber, OP. Preceding Miriam's death, Sister Sharon wrote the following tribute: "I know Miriam to be a woman of faith who understands that faith shows itself in the everyday activities of the individual. I know Miriam to be a Dominican teacher who pursues truth as it exists in the natural world, in our society, and in the individual student with whom she interacts. I know Miriam to be a mentor who believes that the selfless preparation of leaders for the next generation may be the most important task of the leaders of this generation. I know Miriam to be a woman who empowers women, whose students have believed in their own power and acted on that belief because Miriam believed in them."[10] A few days after suffering a massive stroke, Miriam quietly passed away on June 15, 2002.

The development of the KBr disk technique and its application to the study of the DNA bases was the high point of Miriam's scientific career. Her last research article was published in 1954, only two years after her KBr paper. While some scientists, like Watson and Crick, made their biggest discoveries early in their research careers and spent the rest of their days in its shadows, Miriam's breakthrough came triumphantly at the end of hers. In the end, Miriam's gain was not monetary, for she did not patent the KBr disk technique. Instead, Miriam's reward was in the knowledge that her work had brought her closer to God.

Guided by her belief in scientific research as an act of worship, Miriam weaved her way through the DNA labyrinth. Despite personal tragedy and professional hardships, she succeeded in reaching its center. There, Miriam saw the "soul" of DNA through her glass-like disks of KBr.

Chapter 11

Memories

"When she (Miriam) was about 12, she told her sister, Alice Ruth, that since she slept near the window, she was getting ALL the oxygen!"

– Sister Joan Sustersic, OP.

"In 1964 she did a forward-looking thing that aroused criticism. She had a Teilhard de Chardin symposium. She was criticized for teaching evolution. She was interested in helping teachers, and developed science texts for junior high."

– Sister Mary Beaubien, OP.

"She was very concerned about those of us teaching high school science. She made a big effort to provide programs at Siena Heights that would assist us. If you taught in a small high school, you taught all the science courses, and perhaps advanced math. That took a broad education. So she made provisions for us to learn literally everything. Another area that we had to deal with was the concept of the origin of the species. She made sure that we understood the evolutionary processes. She has left her mark on many of us."

– Sister Virginia O'Reilly, OP.

"She was my chemistry teacher in 1957. I remember she always said English should always be 'precise' and 'concise.' When I came back to Siena Heights for a reunion in 1965 and saw Sister Miriam without her habit, I asked her if she wanted to be called Dr. Stimson. She said, 'Just call me Miriam.'"

– Carol (Robel) Krawczak

"She has been telling it like it is for a long time. Too bad we didn't always listen. I'm sure she irritated many along the way with her forward thinking and strong opinions. Thank God for her persistence in the face of disapproval."

– Sister Ursula Ording, OP.

"She was a very colorful person, who was ahead of her time in many ways. Despite her vision problems, she was extremely well read … and passed the materials on … She was very personable – at times brusque, but warm – and she was wise."

– Sister Joan Sustersic, OP.

"… we went to a play. It was a musical on Coco Chanel, and Katharine Hepburn played the part. [Miriam sat next to Sister Nadine Foley, OP, while Katharine Hepburn was on stage] And I thought, 'there's a real similarity to this woman sitting next to me and that woman down on the stage!'"

– Sister Nadine Foley, OP.

"When I was trying to get a visa, my country was taken over by the Communists, so I couldn't get one. Relations between China and the United States became very poor. After many years, relations were improved, and I began to think about taking advantage of the scholarship that had been given to me in 1947. It was still dangerous to write to anyone in the United States – you might be persecuted if you did – but I tried. I wrote there and got a letter back from Sister Miriam." "It is through her that my sister and I could come here [in 1979] and finish our graduate studies. We will always cherish her love and kindness."

– Margaret Chi.

"When she came to Maria [Health Center], she sat at a table in the dining room with three other sisters, including me. She was a very humble person. She would ask us for advice. She was so much better educated than we were; we didn't like to give her advice. But she never showed any kind of a sense of superiority over the rest of us who didn't have her scientific knowledge and her high degrees. She was very dear to us. Her death was a real sorrow for us all."

– Sister Majella Gibson, OP.

"Miriam is at rest, but I imagine that heaven is astir and will never be the same!"

– Sister Marcine Klemm, OP.

Acknowledgements

First and foremost, I am deeply grateful for the cooperation of Sister Miriam Michael Stimson, OP, who graciously consented to numerous interviews as well as provided access to her personal documents.

Thank you, Chandra Elliot, for your invaluable assistance in collecting and organizing the materials from the archives and library of Siena Heights University, the archives of the Adrian Dominican Sisters, the archives of St. Joseph Academy, and the Adrian Public Library.

I also thank the archivists and librarians of Siena Heights University and the Adrian Dominican Sisters: Sister Helen Duggan, OP, Sister Terese Haggarty, OP, and Melissa Sissen.

Thanks, also, to Catherine Dymond for consenting to an interview and for providing photographs of Miriam.

I owe much gratitude to the following for providing wonderful descriptions and anecdotes of Miriam: Christine Kazen Cauchi, Sister Helen Duggan, OP, Dr. Karen Erickson, Sister Nadine Foley, OP, Sister Therese Mary Foote, OP, Sister Irma Gerber, OP, Sister Majella Gibson, OP, Louise Godzina, Joanne Hettrick, Carol (Robel) Krawczak, Diana Albera Luciani, Sister Mary Bernard Lynch, OP, Sister Helen McCartney, OP, Dorothy Nyhoff, Sister Marie Joannes O'Donnell, OP, Julia Franko Opalek, Sister Ursula Ording, OP, Judith Redwine, Jeanne Sheck, Sister Jodie Screes, OP, Sister Marie Siena, OP, Sister Mary Stuart, OP, Sister Anna Sullivan, OP, Dr. Donita Sullivan, Sister Sharon Weber, OP, and Sister Rose Louise Zimmer, OP.

Special thanks to the following for reviewing parts of the manuscript: Sister Nadine Foley, OP, Sister Sharon Weber, OP, Sister Helen Duggan, OP, Catherine Dymond, Sister Majella Gibson, OP, and Sister Marie Joannes O'Donnell, OP.

I am grateful for the support of the Siena Heights University Advancement Office, Jennifer Hamlin Church and Sue Gerhard, for their help in contacting alumni.

Thank you, Ray Casey, Emma McFaul, and Gabrielle Clemons for your technical assistance in scanning and editing photographs. I also thank Steve Wathen for helping me to draw the chemical structures.

I also thank Biologics International, Inc. (Shara Tackett) and John Heitmann (University of Dayton) for the literature on Dr. Sperti, the Institutum Divi Thomae, and the biodyne research.

This project was supported by a 4-month-long sabbatical from Siena Heights University, as well as by a grant from the Chemical Heritage Foundation.

Credits

The photographs that appear in this book are reproduced from the following sources:

1. Siena Heights University. Cover photograph of Sister Miriam Michael Stimson, OP.

2. Catherine Dymond. Photograph of Marian (Miriam) and Alice Ruth Stimson.

3. Siena Heights University. Marian (Miriam) Stimson's high school graduation picture.

4. Siena Heights University. Sister Miriam Michael Stimson, OP, and Dr. Elton Cook.

5. *Action*. The 1939 graduates of the Institutum Divi Thomae.

6. *Cincinnati Times-Star*. Sister Miriam Michael Stimson, OP, using a Hilger quartz spectrograph in conjunction with a Spekker photometer.

7. Siena Heights University. Sister Miriam Michael Stimson, OP, with a galvanometer.

8. Adrian Dominican Sisters. Sister Miriam Michael Stimson, OP, adjusting the pH of her solution.

9.   *Adrian Daily Telegram.* Sister Miriam Michael Stimson, OP, with her press and die.

10. Adrian Dominican Sisters. Sister Miriam Michael Stimson, OP, with her mother and Pope Pius XII.

Notes

Before her death, Miriam generously gave me access to her personal scrapbooks, which included a wealth of newspaper and magazine clippings. Some of these clippings do not contain complete publishing information, and I have indicated these incompletely annotated sources by the abbreviation *IAS*.

Chapter 2. A Disease of Our Genes

1.   Patterson, James. *The Dread Disease: Cancer and Modern American Culture*. Cambridge, MA: Harvard University Press, 1987, p. 30.

2.   Ibid, p. 29.

3.   Ibid, p. 12.

4.   Ibid, pp. 19-20.

5.   Ibid, p. 24.

6.   Koch, William. *Cancer and its Allied Diseases*. Detroit, MI, 1929, p. 33.

7.   American Cancer Society. *Cancer Facts & Figures 2002*. ACS, Atlanta, GA: ACS, 2002, p. 4.

8.   Patterson, James. *The Dread Disease: Cancer and Modern American Culture*. Cambridge, MA: Harvard University Press, 1987, p. 59-60.

9.   Ibid, p. 62.

10.   American Cancer Society. *Cancer Facts & Figures 2002*. ACS, Atlanta, GA: ACS, 2002.

11.   Lichtenstein, Paul., Niels Holm, Pia Verkasalo, Anastasia Ili-adou, Jaakko Kiaprio, Markku Koskenvuo, Eero Pukkala, Axel Skytthe, and Kari Hemminki. "Environmental and Heritable Factors in the Causation of Cancer." *The New England Journal of Medicine* 343 (2000): 78-85.

Chapter 3. The Roots of Humanity

1.   Bronowski, Jacob. *The Ascent of Man*. Boston, MA: Little, Brown and Company, 1973, p. 64.

2.   Keller, Evelyn Fox. *The Century of the Gene*. Cambridge, MA: Harvard University Press, 2000, p. 2.

3.   Keller, Evelyn Fox. *A Feeling for the Organism: The Life and Work of Barbara McClintock*. San Francisco: W. H. Freeman, 1983, p. 97.

4.   Weaver, Robert, and Philip Hedrick. *Basic Genetics*. Dubuque, IA: Wm C. Brown Publishers, 1995, p. 75.

Chapter 4. Fibers of Life

1.   Schrödinger, Erwin. *What is Life? The Physical Aspect of the Living Cell*. New York: MacMillan Co., 1945, pp. 3 & 61.

2.   Ibid, p. 61.

3.   Ibid, p. 86.

4.   Watson, James. 1968. *The Double Helix: A Personal Account of the Discovery of the Structure of DNA*. New York: Atheneum Publishers, 1968, p. 112.

Chapter 5. A Bride of Christ

1.   Stimson, Sister Miriam Michael, OP. Personal interview, June 4, 2002.

2.   Ibid.

3.   Ibid.

4.   Dymond, Catherine. Personal interview, March 14, 2002.

5.   Ibid.

6.   Stimson, Sister Miriam Michael, OP. Personal interview. November 29, 2001.

7.   Dominican Finds Way to Serve God in Science. FIAT, Volume I, No. 2, 1955.

8.   Stimson, Sister Miriam Michael, OP. Personal interview, December, 13, 2001.

9.   Ryan, Sister Mary Philip, OP. *Amid the Alien Corn*. St. Charles, Illinois: Jones Wood Press, 1967, p. 127.

10. Ibid.

11. Ryan, op. cit., p. 191.

12. Ryan, op. cit., p. 198.

13. Ryan, op. cit., p. 200.

14. Ryan, op. cit., p. 208.

15. Stimson, Sister Miriam Michael, OP. Personal interview, November 9, 2001.

16. Neckel, Louise; Schulte, Mary; and Flower, Irene. Golden Jubilee Class '32. Saint Joseph Academy, Adrian, Michigan, Alumnae Reunion. October, 1982.

17.   Ryan, op. cit., p. 302.

18.   Ibid.

19. Zimmer, Sister Rose Louise, OP. Personal correspondence, 2002.

20.   Godzina, Louise (Neckel). Personal correspondence, 2002.

21. Stimson, Sister Miriam Michael, OP. Personal interview, October 25, 2001.

22.   Ibid.

23.   Ibid.

24.   Heidbreder, Jo. "We visit with Dean Miriam Stimson." *Blissfield Advance*. November 17, 1982.

25.   Ryan, op. cit., pp. 293-4.

26.   Ryan, op. cit., p. 297.

27.   Wong, Mary Gilligan. *Nun*. New York: Harcourt Brace Jovanovich, 1983, p. 26; *Maidens of Hallowed Names*. 5 Barclay Street: Excelsior Catholic Publishing House, 1893, pp. 1-11.

28.   Noffke, Suzanne, OP. 1980. *Catherine of Siena: The Dialogue*. New York: Paulist Press, pp. 3-7; The Dominican House of Studies. *Dominican Saints*. The Rosary Press, 1921, pp. 218-229).

29.   Stimson, Sister Miriam Michael, OP. Personal interview conducted by Sister Helen Duggan, OP, July 26, 1991.

30.   Ibid.

31.   Ibid.

32.   Stimson, Sister Miriam Michael, OP. Personal interview. November 9, 2001.

33. Stimson, Sister Miriam Michael, OP. Personal interview. November 29, 2001.

34. Podvin, Sister Catherine, OP. 2002. Adrian Dominican Sisters Archives summary of Miriam Stimson.

35. Wong, op. cit., p. 27.

36. Dymond, Catherine. Personal interview, March 14, 2002.

37. Ibid.

38. Stimson, Sister Miriam Michael, OP. Personal interview, November 2, 2001.

39. Gibson, Sister Majella, OP. Personal interview. February 11, 2002.

40. Ibid.

41. Stimson, Sister Miriam Michael, OP. Personal interview conducted by Sister Helen Duggan, OP, July 26, 1991.

42. Gibson, Sister Majella, OP. Personal interview. February 11, 2002.

43. Wong, op. cit., pp. 202-3.

44. Gibson, Sister Majella, OP. Personal interview. February 11, 2002.

45. Ibid.

46. Podvin, Sister Catherine, OP, op. cit.

47. Gibson, Sister Majella, OP. Personal interview. February 11, 2002.

48. Adrian Dominican Sisters. 2002. Available from World Wide Web: (http://www.adriansisters.org/).

49. Stimson, Sister Miriam Michael, OP. Personal interview. November 16, 2001.

Chapter 6. An Act of Worship

1. Foley, Sisters Nadine, OP, and Mary Philip Ryan, OP. *Mother Mary Gerald Barry, O.P.: Ecclesial Woman of Vision and Daring*. Warren, OH: Superior Printing, 2000, p. 23.

2. Stimson, Sister Miriam Michael, OP. Interviewed by Sister Helen Duggan, OP, August 16, 1991.

3. Wong, op. cit., pp. 245-6.

4. "Sister Mary Jane Hart." Adrian Dominican Sisters, 1980.

5. Elgin, Kathleen. *NUN: A Gallery of Sisters*. New York: Random House, 1964.

6. "Wealth of chemistry experience." *IAS*.

7. Stimson, Sister Miriam Michael, OP. Personal interview, February 26, 2002.

8. McKay, Robert. "The Amazing Dr. Sperti." *Cincinnati*. August 1982, pp. 46-52.

9. Miller, Lois Mattox. "Biodynes." *Scientific American*. January 1943, pp. 14-16.

10. McKay, op. cit., pp. 49 & 52.

11. "Genius in the Attic." *Cincinnati Alumnus*. July, 1968, pp. 6-11.

12. Miller, op. cit., p. 16.

13. "Genius in the Attic," op. cit., p. 9.

14. Ibid.

15. Detzel, Helen. "Institutum Divi Thomae Making Cincinnati Center for Scientific Research." *The Cincinnati Times-Star*. October 22, 1940, p. 22.

16. Ibid.

17. Stimson, Sister Miriam Michael, OP. Personal interview, November 9, 2001.

18. McKay, op. cit., p. 48.

19. "Genius in the Attic," op. cit., p. 9.

20. Ibid.

21. McKay, op. cit., pp. 47 & 48.

22. Detzel, op. cit., p. 22.

23. "A Medieval School in a Modern Setting." *Action: A Catholic Pictorial News Monthly* 2 (1939): 5 – 10.

24. Ibid.

25. Stimson, Sister Miriam Michael, OP. Personal interview, November 2, 2001.

26. Heitmann, John. "Doing 'True Science': The Early History of the Institutum Divi Thomae, 1935-1951." *The Catholic Historical Review* LXXXVIII (2002): 702-722.

27. "A Medieval School in a Modern Setting," op. cit., p. 10.

28. Heitmann, op. cit., p. 714.
29. Elgin, op. cit.

30. Heitmann, op. cit., p. 706.

31. Stimson, Sister Miriam Michael, OP. "Adrian Dominican Roots." *SHU: Reaching New Heights.* 2002.

32. Weber, Sister Sharon, OP. March 20, 2002 & June 19, 2002.

33. "Dominican Finds Way to Serve God in Science." *FIAT* I (1955): spring.
34. "A Medieval School in a Modern Setting," op. cit., p. 5.

35. Stimson, Sister Miriam Michael, OP. Interviewed by Sister Helen Duggan, OP, July 26, 1991.

36.   Ibid.

37.   Stimson, Sister Miriam Michael, OP. Personal interview, November 11, 2001.

38.   Ibid.

39.   Ibid.

40.   McKay, op. cit., pp. 46-52.

41.   Ibid.

42.   "Genius in the Attic," op. cit., p. 8; Miller, op. cit., p. 14.

43.   Ibid.

44.   "Genius in the Attic," op. cit., p. 8.

45.   Fardon, John, Robert Norris, John Loofbourow, and Sister M. Veronita Ruddy, O.P. "Stimulating Materials obtained from Injured and Killed Cells." *Nature* 139 (1937): 589. Sperti, George, John Loofbourow, and Sister Mary Michaella Lane, S. C. "Effects on Tissue Cultures of Inter-Cellular Hormones from Injured Cells." *Science* 86 (1937): 611. Sperti, George, John Loofbourow, and Cecilia Dwyer. "Proliferation-promoting Substances from Cells injured by Ultra-violet Radiation." *Nature* 140 (1939): 643-644.

46.   Fardon, op. cit., p.589.

47.   "Genius in the Attic," op. cit., p. 8.

48.   Miller, op. cit., pp. 14-16.

49.   "Genius in the Attic," op. cit., p. 8; Miller, op. cit., pp. 14-16.
50.   Gariepy, Jennifer. "Stimson to Retire." *Spectra*. January 23, 1991.

51.   Applegate, Robert. "Dominican Sister describes work in cancer research." *Adrian Daily Telegram*. April 11, 1951.

52.   Russell, Margaret. "Adrian Sister's Science Aid in War on Cancer." *Detroit Sunday Times*. June 14, 1953.

53.   Loofbourow, John R., Elton S. Cook, and Sister Miriam Michael Stimson. "Chemical Nature of Proliferation-Promoting Factors from Injured Cells." *Nature* 142 (1938): 573.

54.   Loofbourow, John R., Elton S. Cook, and Sister Miriam Michael Stimson. "Chemical Nature of Proliferation-Promoting Factors From Injured Cells." *Nature* 142 (1938): 573. Cook, Elton S., John R. Loofbourow, and Sister Miriam Michael Stimson, O.P. "Chemical Studies of Proliferation-promoting Factors From Ultra-violet Injured Cells." *Atti X^o congr. Intern. Chim.* 5 (1939): 26-34.

55.   Cook, Elton S., Sister Mary Jane Hart, O.P., and Sister Miriam Michael Stimson, O.P. "Proliferation-promoting Properties and Ultraviolet Absorption Spectra of Fractions From Yeast." *The Biochemical Journal* 34 (1940): 1580-1587.

56. Stimson, Sister Miriam Michael, OP. Interviewed by Sister Helen Duggan, OP, July 26, 1991.

57.   Loofbourow, John R. and Miriam M. Stimson. "Ultra-violet Absorption Spectra of Nitrogenous Heterocyclic Compounds. Part I. Effects of pH and Irradiation on the Spectrum of Adenine." *Journal of the Chemical Society* (1940): 844-848. Loofbourow, John R. and Miriam M. Stimson. "Ultra-violet Absorption Spectra of Nitrogenous Heterocyclic Compounds. Part II. Effects of pH and Irradiation on the Spectrum of Barbituric Acid." *Journal of the Chemical Society* (1940): 1275-1277.

58.   Cook, Elton S., Sister Mary Jane Hart, O.P., and Sister Miriam Michael Stimson, O.P. "Proliferation-promoting Properties and Ultraviolet Absorption Spectra of Fractions From Yeast." *The Biochemical Journal* 34 (1940): 1580-1587.

59.   Bentley, J. Peter, Thomas Hunt, Jacqueline Weiss, Christopher Taylor, Albert Hanson, Gordan Davies, and Betty Halliday. "Peptides From Live Yeast Cell Derivative Stimulate Wound Healing." *Archives of Surgery* 125 (1990): 641-646.

60.   "Sister Mary Jane Hart," op. cit.

61.   Panian, David. "Sister Miriam Stimson debunks 'urban legend' of Preparation H patent." *Adrian Daily Telegram*. April 10, 2000.

Chapter 7. Miriam's Children

1.   "Sister Grace Reuter." Adrian Dominican Sisters, 1986.

2.   Ibid.

3.   Haggerty, Sister Therese, OP. Personal communication, 2003.

4.   "Sister Grace Reuter," op. cit.

5.   Luciani, Diana Albera. Personal correspondence, 1993.

6.   Luciani, Diana Albera. Personal correspondence, May 24, 1993.

7.   Luciani, Diana Albera. Personal correspondence, March 2002.

8.   Luciani, Diana Albera. Personal correspondence, July 6, 1993.

9.   Ibid.

10.   Luciani, Diana Albera. Personal correspondence, March 2002.

11.   Sheck, Jeanne. Personal correspondence, 2002.

12.   Stimson, Sister Miriam Michael, OP. Interview by Sister Helen Duggan, OP, July 26, 1991.

13.   Ibid.

14. Stimson, Sister Miriam Michael, OP. Personal interview, December 13, 2001.

15. Stimson, Sister Miriam Michael, OP. Interview by Sister Helen Duggan, OP, July 26, 1991.

16. Ibid.

17. Applegate, Robert. "Dominican Sister describes work in cancer research." *Adrian Daily Telegram.* April 11, 1951.

18. Stimson, Sister Miriam Michael, OP. Personal interview, November 2, 2001.

19. Gariepy, Jennifer. "Stimson to Retire." *Spectra.* January 23, 1991.

20. Stimson, Sister Miriam Michael, OP. *Molecular Structure of Nitrogenous Heterocycles.* Cincinnati: Institutum Divi Thomae, 1948.

21. Loofbourow, John R. and Miriam Michael Stimson. "Ultra-violet Absorption Spectra of Nitrogenous Heterocyclic Compounds. Part I. Effects of pH and Irradiation on the Spectrum of Adenine." *Journal of the Chemical Society* (1940): 844-848.

22. Loofbourow, John R., Sister Miriam Michael Stimson, and Sister Mary Jane Hart. "The Ultraviolet Absorption Spectra of Nitrogenous Heterocycles. V. The Blocking Effect of Methyl Groups on the Ultraviolet Absorption Spectra of Some Hydroxypurines and Pyrimidines." *Journal of the American Chemical Society* 65 (1943): 148-151.

23. Stimson, Sister Miriam Michael and Sister Mary Agnita Reuter. "The Effect of pH on the Spectra of Thymine and Thymine Desoxyriboside." *Journal of the American Chemical Society* 67 (1945): 847-848.

24. Stimson, Sister Miriam Michael and Sister Mary Agnita Reuter. "The Ultraviolet Absorption Spectra of Cytosine and Isocytosine." *Journal of the American Chemical Society* 67 (1945): 2191-2193.

25. Stimson, Sister Miriam Michael, O.P. *Molecular Structure of Nitrogenous Heterocycles*. Cincinnati: Institutum Divi Thomae, 1948.

26. Pazdera, H. J. and W. H. McMullen. "Chromatography: Paper." *Treatise on Analytical Chemistry*. Part I, Theory and Practice, Volume 3. Edited by I. M. Kolthoff and P. J. Elving. New York: Interscience Publishers, 1961.

27. Stimson, Sister Miriam Michael, OP. Interview by Sister Helen Duggan, OP, July 26, 1991.

28. Stimson, Sister Miriam Michael, OP. Personal interview, February 26, 2002.

29. Stimson, Sister Miriam Michael and Sister Mary Agnita Reuter. "The Fluorescence of Some Purines and Pyrimidines." *Journal of the American Chemical Society* 63 (1941): 697-699.

30. "Drinkers of Whisky Warned of Water-Mix with Liquor." *IAS*.

31. Chargaff, Erwin. *Heraclitean Fire: Sketches from a Life before Nature*. New York: Rockefeller University Press, 1978, pp. 83-84.

32. Olby, Robert. *The Path to the Double Helix*. Seattle: University of Washington Press, 1974, p. 87.

33. Vischer, Ernst and Erwin Chargaff. "The Separation and Characterization of Purines in Minute Amounts of Nucleic Acid Hydrolysates." *Journal of Biological Chemistry* 168 (1947): 781-782.

34. Olby, op. cit., p. 212.

35. Chargaff, Erwin. *Essays on Nucleic Acids*. New York: Elsevier Publishing Co., 1963, 9.

36. Vischer, Ernst and Erwin Chargaff. "The Separation and Quantitative Estimation of Purines and Pyrimidines in Minute Amounts." *Journal of Biological Chemistry* 176 (1948): 703-714.

37. Kalckar, Herman. M. "Differential Spectrophotometry of Purine Compounds by Means of Specific Enzymes. I. Determination of Hydroxypurine Compounds." *Journal of Biological Chemistry* 167 (1947): 429-443.

38. Chargaff, Erwin. "Base Composition of Deoxypentose and Pentose Nucleic Acids in Various Species." *The Chemical Basis of Heredity*. Edited by William McElroy and Bentley Glass. Baltimore: John Hopkins Press, 1954.

39. Chargaff, Erwin. *Essays on Nucleic Acids*. New York: Elsevier Publishing Co., 1963, p. 21.

40. Olby, op.cit., p. 214.

Chapter 8. Tautomerism

1. Stimson, Miriam Michael and Mary Agnita Reuter. "The Spectrophotometric Estimation of Methoxy-cinchona Alkaloids." *Journal of the American Chemical Society* 68 (1946): 1192-1196.

2. Stimson, Sister Miriam Michael, O.P. Interviewed by Sister Helen Duggan, OP. July 26, 1991.

3. Stimson, Sister Miriam Michael and Sister Mary Agnita Reuter. "The Effect of pH on the Spectra of Thymine and Thymine Desoxyriboside." *Journal of the American Chemical Society* 67 (1945): 847-848.

4. Loofbourow, John and Miriam M. Stimson. "Ultra-violet Absorption Spectra of Nitrogenous Heterocyclic Compounds. Part I. Effect of pH and Irradiation on the Spectrum of Adenine." *Journal of the Chemical Society* (1940): 844-848.

5. Loofbourow, John R., Sister Miriam Michael Stimson, and Sister Mary Jane Hart. "The Ultraviolet Absorption Spectra of Nitrogenous Heterocycles. V. The Blocking Effect of Methyl Groups on the Ultraviolet Absorption Spectra of Some Hydroxypurines and Pyrimidines." *Journal of the American Chemical Society* 65 (1943): 148-151.

6. Stimson, Sister Miriam Michael and Sister Mary Agnita Reuter. "The Ultraviolet Absorption Spectra of Cytosine and Isocytosine." *Journal of the American Chemical Society* 67 (1945): 2191-2193.

7. Ibid.

8. Heyroth, Francis and John Loofbourow. "Changes in the Ultraviolet Absorption Spectrum of Uracil and Related Compounds Under the Influence of Radiations." *Journal of the American Chemical Society* 53 (1931): 3441-3452.

9. Loofbourow, John and Miriam M. Stimson. "Ultra-violet Absorption Spectra of Nitrogenous Heterocyclic Compounds. Part II. Effect of pH and Irradiation on the Spectrum of Barbituric Acid." *Journal of the Chemical Society* (1940): 1275-1277.

10. Loofbourow, John R., Sister Miriam Michael Stimson, and Sister Mary Jane Hart. "The Ultraviolet Absorption Spectra of Nitrogenous Heterocycles. V. The Blocking Effect of Methyl Groups on the Ultraviolet Absorption Spectra of Some Hydroxypurines and Pyrimidines." *Journal of the American Chemical Society* 65 (1943): 148-151.

11. Loofbourow, John and Miriam M. Stimson. "Ultra-violet Absorption Spectra of Nitrogenous Heterocyclic Compounds. Part I. Effect of pH and Irradiation on the Spectrum of Adenine." *Journal of the Chemical Society* (1940): 844-848.

12. Stimson, Sister Miriam Michael, O.P. *Molecular Structure of Nitrogenous Heterocycles.* Cincinnati: Institutum Divi Thomae, 1948.

13. Stimson, Sister Miriam Michael. "The Ultraviolet Absorption Spectra of Some Pyrimidines. Chemical Structure and the Effect of pH on the Position of lambda max." *Journal of the American Chemical Society* 71 (1949): 1470-1474.

14. Marshall, J. R. and James Walker. "An Experimental Study of Some Potentially Tautomeric 2- and 4(6)-Substituted Pyrimidines." *Journal of the Chemical Society* 224 (1951): 1004-1017.

15. Heyroth, Francis and John Loofbourow. "Correlation of Ultraviolet Absorption and Chemical Constitution in Various Pyrimidines and Purines." *Journal of the American Chemical Society* 56 (1934): 1728-1734. Cavalieri, Liebe, Aaron Bendich, John Tinker, and Beorge Brown. "Ultraviolet Absorption Spectra of Purines, Pyrimidines and Triazolopyrimidines." *Journal of the American Chemical Society* 70 (1948): 3875-3880. Sinsheimer, R., J. Scott, and J. Loofbourow. "Ultraviolet Absorption Spectra at Reduced Temperatures. II. Pyrimidines and Purines." *Journal of Biological Chemistry* 187 (1950): 313-324.

16. Stimson, Miriam Michael and Mary Agnita Reuter. "The Spectrophotometric Estimation of Methoxy-cinchona Alkaloids." *Journal of the American Chemical Society* 68 (1946): 1192-1196.

17. Pauling, Linus and Robert Corey. "Specific Hydrogen-Bond Formation Between Pyrimidines and Purines in Deoxyribonucleic Acids." *Archives of Biochemistry and Biophysics* 65 (1956): 164-181.

18. Olby, op. cit., p. 392.

19. Olby, op. cit., p. 393.

20. Olby, op. cit., p. 360.

21. Watson, James. *The Double Helix: A Personal Account of the Discovery of the Structure of DNA*. New York: Atheneum Publishers, 1968, pp. 115-6.

22. Pauling, Linus and Robert Corey. "A Proposed Structure for the Nucleic Acids." *Proceedings of the National Academy of Sciences, U.S.A.* 39 (1953): 84-97.

23. Watson, op. cit., p. 108.

24. Crick, Francis. *What Mad Pursuit: A Personal View of Scientific Discovery*. New York: Basic Books, 1988, p. 70.

25. Watson, op. cit., p. 112.

26. Ibid.

27. Olby, op. cit., p. 405.

28. Olby, op. cit., p. 389.

29. Chargaff, Erwin. *Heraclitean Fire: Sketches from a Life before Nature.* New York: Rockefeller University Press, 1978, pp. 101-2.

30. McElheny, Victor. *Watson and DNA: Making a Scientific Revolution.* New York: Perseus Publishing, 2003, p. 48.

31. Davidson, J. N. *The Biochemistry of the Nucleic Acids.* London: Methuen & Co., Ltd., 1953.

32. Watson, op. cit., p. 115.

33. Gulland, J., D. Jordan, and W. Taylor. "Deoxypentose Nucleic Acids. Part II. Electrometric Titration of the Acidic and the Basic Groups of the Deoxypentose Nucleic Acid of Calf Thymus." *Journal of the Chemical Society* (1947): 1131-1141.

34. Watson, op. cit., p. 116.

35. Watson, op. cit., pp. 117-8.

36. Watson, op. cit., pp. 120-122.

37. Olby, op. cit., p. 410.

38. Watson, op. cit., pp. 122-123.

39. Watson, op. cit., pp. 123-125.

40. Crick, op. cit., p. 71.

41. Watson, James and Francis Crick. "Molecular Structure of Nucleic Acids: A Structure for Deoxyribose Nucleic Acid." *Nature* 171 (1953): 737-738.

42. Wilkins, Maurice, Alex Stokes, and H. R. Wilson. "Molecular Structure of Deoxypentose Nucleic Acids." *Nature* 171 (1953): 738-740.

43. Franklin, Rosalind and Raymond Gosling. "Molecular Configuration in Sodium Thymonucleate." *Nature* 171 (1953): 740-741.

44. Chargaff, Erwin. *Essays on Nucleic Acids*. New York: Elsevier Publishing Co., 1963, p. 53.

45. Judson, Horace. 1979. *The Eighth Day of Creation: Makers of the Revolution in Biology*. New York: Simon and Schuster, 1979, p. 186.

Chapter 9. KBr

1. Donohue, Jerry. "Hydrogen-bonded Helical Configurations of Polynucleotides." *Proceedings of the National Academy of Sciences, U.S.A.* 42 (1956): 60-65.

2. Pauling, Linus, and Robert Corey. "Specific Hydrogen-Bond Formation between Pyrimidines and Purines in Deoxyribonucleic Acids." *Archives of Biochemistry and Biophysics* 65 (1956): 164-181.

3. Ulbricht, T. L. V. *Purines, Pyrimidines and Nucleotides and the Chemistry of Nucleic Acids*. London: Pergamon Press, 1964.

4. Heitmann, op. cit.

5. Gariepy, op. cit.

6. Stimson, Sister Miriam Michael, O.P. Personal interview, February 26, 2002.

7. Nyhoff (O'Kane), Dorothy. 2002. Personal letter.

8. Luciani, Diana Albera. July 6, 1993.

9. Luciani, Diana Albera. Personal letter, March, 2002.

10. Watson, op. cit., p. 143.

11. Gariepy, op. cit.

12. Luciani, Diana Albera. Personal letter, March, 2002.

13. Stimson, Sister Miriam Michael, OP. Personal interview conducted by Sister Helen Duggan, OP, July 26, 1991.

14. Luciani, Diana Albera. Personal letter, March 2002.

15. Watson, op. cit., p. 21.

16. "Dominican finds way to serve God in science." *FIAT* 1 (1956): spring.

17. Stimson, Sister Miriam, O.P. 1959. "The use of spectral absorption in chemistry and biology." *The Science Counselor* XXII (1959): winter.

18. Smith, A. Lee. "Infrared Spectroscopy." *Treatise on Analytical Chemistry*. Part I, Volume 6. Edited by M. Kolthoff and Philip Elving. New York: Interscience Publishers, 1965, p. 3539.

19. Elgin, op. cit., pp. 22-25.

20. Smith, A. Lee. "Infrared Spectroscopy." *Treatise on Analytical Chemistry*. Part I, Volume 6. Edited by M. Kolthoff and Philip Elving. New York: Interscience Publishers, 1965, p. 3539.

21. Elgin, op. cit., pp. 22-25.

22. Harrison, George, Richard Lord, and John Loofbourow. *Practical Spectroscopy*. New York: Prentice-Hall, 1948, p. 503.

23. Chemical Heritage Foundation. Museum exhibit, May, 2003.

24. Beaven, G. H., E. R. Holliday, and E. A. Johnson. "Optical Properties of Nucleic Acids and Their Components." *The Nucleic Acids: Chemistry and Biology*. Edited by Erwin Chargaff and J.

N. Davidson. New York: Academic Press, Inc., 1955, pp. 546-547.

25. Stimson, Sister Miriam, O.P. "Sampling for Spectrophotometry with Special Reference to the Use of the Potassium Bromide Disk Technique." *Progress in Infrared Spectroscopy*. New York: Plenum Press, 1962, pp. 143-150.

26. "Sister Grace Reuter," op. cit.

27. O'Donnell, Sister Marie Joannes, OP. Personal interview, March, 19, 2002.

28. Luciani, Diana Albera, May 24, 1993.

29. Harrison, George, Richard Lord, and John Loofbourow, op. cit., p. 503.

30. Sinsheimer, R., J. Scott, and J. Loofbourow. "Ultraviolet Absorption Spectra at Reduced Temperatures. II. Pyrimidines and Purines." *Journal of Biological Chemistry* 187 (1950): 313-324.

31. Cook, Elton, Cornelius Kreke, Edward Barnes, and Werner Motzel. "Infra-red and Ultra-violet Absorption Spectra of Proteins in the Solid State." *Nature* 174 (1954): 1144-1147.

32. Smith, op. cit., p. 3589.

33. Fridmann, Sherrill. "Pelleting Techniques in Infrared Analysis – A Review and Evaluation." *Progress in Infrared Spectroscopy*. Volume 3. Edited by Herman Szymanski. New York: Plenum Press, 1966.

34. Stimson, Sister Miriam Michael, OP. Personal interview, October, 25, 2001.

35. Stimson, Sister Miriam Michael, OP. Personal interview, February 26, 2002.

36. Ibid.

37. O'Donnell, Sister Marie Joannes, OP. Personal interview, March, 19, 2002.

38. Elgin, op. cit., pp. 22-25.

39. O'Donnell, Sister Marie Joannes, OP. Personal interview, March, 19, 2002.

40. Applegate, Robert. "Dominican Sister Describes Work in Cancer Research." *Adrian Daily Telegram*, April 11, 1951.

41. Ibid.

42. Elgin, op. cit., pp. 22-25.

43. O'Donnell, Sister Marie Joannes, OP. 2002. Personal interview.

44. Ibid.

45. Luciani, Diana Albera, 1993.

46. O'Donnell, Sister Marie Joannes, OP. Personal interview, March, 19, 2002.

47. "New Equipment for Research." *Adrian Daily Telegram*, April 8, 1952.

48. Stimson, Sister Miriam Michael, OP. Personal interview, November 2, 2001.

49. Erickson, Karen. Personal interview, March 9, 2004.

50. Stimson, Sister Miriam Michael, OP. Personal interview, February 26, 2002.

51. O'Donnell, Sister Marie Joannes, OP. Personal interview, March, 19, 2002.

52. Rao, C. N. R. *Chemical Applications of Infrared Spectroscopy*. New York: Academic Press, 1963, p. 68.

53. Kendall, David. "Sample Preparation Procedures." *Applied Infrared Spectroscopy*. Edited by David Kendall. New York: Reinhold Publishing Corp., 1966, p. 145.

54. Elgin, op. cit., pp. 22-25.

55. Stimson, Sister Miriam Michael and Sister Marie Joannes O'Donnell. "The Infrared and Ultraviolet Absorption Spectra of Cytosine and Isocytosine in the Solid State." *Journal of the American Chemical Society* 74 (1952): 1805-1808.

56. "Siena Scientist is Conference Speaker." *Adrian Daily Telegram*, September 5, 1951.

57. Elgin, op. cit., pp. 22-25.

58. Stimson, Sister Miriam Michael, OP. Personal interview, February 26, 2002.

59. Stimson, Sister Miriam Michael, OP. Personal interview, November 2, 2001.

60. Stimson, Sister Miriam Michael, OP. Personal interview, November 29, 2001.

61. Ibid.

62. Stimson, Sister Miriam Michael, OP. Personal interview, February 26, 2002.

63. Rose, Judith. Dominican Nun Pioneers Cell Research. *IAS.*

64. Elgin, op. Cit., pp. 22-25.

65. Rose, op. cit.

66. Schiedt, Ulrich, and Helmuth Reinwein. "Zur Infrarot-Spektroskopie von Aminosauren. I. Mitt.: Eine neue Praparationstechnik zur Infrarot-Spektroskopie von Aminosauren und anderen polaren Verbindungen." *Zeitschrift für Naturforschung* 7b (1952): 270-277.

67. Schiedt, Ulrich. "Zur Infrarot-Spektroskopie von Aminosauren. II. Mitte.: Eine verbesserte, zur infrarot-photometrischen Bestimmung von Aminosauren und anderen polaren

Verbindungen geeignete Praparationstechnik." *Zeitschrift für Naturforschung* 8b (1953): 66-70.

68. Rose, op. cit.

69. Elgin, op. cit., pp. 22-25.

70. Stimson, Sister Miriam Michael, OP. Personal interview, November 25, 2001.

71. Stimson, Sister Miriam Michael, OP. Personal interview, December 7, 2001.

72. Stimson, Sister Miriam Michael, O.P. "L'Absorption Infrarouge et Ultraviolette de Divers Heterocycles Azotes Diaminosubstitues." *Journal de Physique et le Radium* 15 (1954): 390-393. Stimson, Sister Miriam Michael, O.P. "Sodium and Potassium Cation Dependence of the Infrared Absorption of COO⁻." *Journal of Chemical Physics* 22 (1954): 1942.

73. "Nun's Work in Chemistry Honored." The Michigan Catholic, February 25, 1954.

74. Beaven, G. H., E. R. Holliday, and E. A. Johnson. "Optical Properties of Nucleic Acids and Their Components." *The Nucleic Acids: Chemistry and Biology*. Edited by Erwin Chargaff and J. N. Davidson. New York: Academic Press, Inc., 1955, pp. 546-547.

75. Hannah, R. and J. Swinehart. *Experiments in Techniques of Infrared Spectroscopy*. Norwalk, CT: The Perkin-Elmer Corporation, 1974, pp. 7-1 to 7-4.

76. Rao, C. N. R. *Chemical Applications of Infrared Spectroscopy*. New York: Academic Press, 1963. Kendall, David. "Sample Preparation Procedures." *Applied Infrared Spectroscopy*. Edited by David Kendall. New York: Reinhold Publishing Corp., 1966. Fridmann, Sherrill. "Pelleting Techniques in Infrared Analysis – A Review and Evaluation." *Progress in Infrared Spectroscopy, Volume 3*. Edited by Herman Szymanski. New York: Plenum Press, 1966.

77.   Gibson, Sister Majella, OP. Personal correspondence, 2002.

78.   Cook, Elton. "Science Has Led to Medicine's Advances." *IAS*.

Chapter 10. Unveiled

1.   Angell, C. "An infrared spectroscopic investigation of nucleic acid constituents." *Journal of the Chemical Society* (1961): 504-515.

2.   Ulbricht, T. L. V. *Purines, Pyrimidines and Nucleotides and the Chemistry of Nucleic Acids*. London: Pergamon Press, 1964, p. 5.

3.   Stent, Gunther. "The Aperiodic Crystal of Heredity." *DNA: The Double Helix: Perspective and Prospective at Forty Years*. Edited by Donald Chambers. New York: The New York Academy of Sciences, 1995, p. 30.

4.   Watson, J. D. and F. H. C. Crick. "A Structure for Deoxyribose Nucleic Acid." *Nature* 171 (1953): 737-8.

5.   Watson, J. D. and F. H. C. Crick. "Genetic Implications of the Structure of Deoxyribonucleic acid." *Nature* 171 (1953): 964-7.

6.   Ibid.

7.   Bowden, Mary Ellen, Amy Beth Crow, and Tracy Sullivan. *Pharmaceutical Achievers: The Human Face of Pharmaceutical Research*. Philadelphia: Chemical Heritage Press, 2003.

8.   Watson, James. "A Personal View of the Project." *The Code of Codes*. Edited by Daniel Kevles and Leroy Hood. Cambridge, MA: Harvard University Press, 1992, p. 166.

9.   Stimson, Sister Miriam Michael, OP. Personal interview conducted by Sister Jodie Screes, OP, November 6, 1996.

10.   Weber, Sister Sharon, OP, January 17, 1991.

Questions for Discussion

1. How was Miriam's life and work affected by her belief in God? How has your life been influenced by your faith?

2. In what manner was Miriam such a memorable teacher? As a student, how would you have responded to such a teacher?

3. What were some of the personal joys and tragedies that Miriam experienced? If you were in her shoes, how would you have responded?

4. Describe the major world events of the 1930s and '40s and how they affected Miriam. How has current world events influenced your life?

5. How did Miriam finance her research? How would you go about raising funds for a project?

6. In what ways was Miriam neither a stereotypical religious sister nor a typical scientist? How do you differ from the norm?

7. What kind of discrimination did Miriam face in her lifetime? What kind of obstacles have you encountered?

8. Describe how the Catholic Church, the Adrian Dominican Sisters, St. Joseph Academy, St. Joseph College, or Siena Heights College functioned as an extended family for Miriam. Describe the importance of "family" in your life.

9.   How did Miriam, and other religious sisters like her, help to re-
       define the roles open for women? How would you define
       woman?

10.   How did Miriam express her affection for art and music? How
       has art and music influenced your life?

Made in the USA
Middletown, DE
27 November 2014